Copyright © 2016 by Lou Giannelli.

Library of Congress Control Number: 2016906461
ISBN: Hardcover 978-1-5144-8712-9
Softcover 978-1-5144-8711-2
eBook 978-1-5144-8709-9

All rights reserved. No part of this book may be reproduced or transmitted in any form or by any means, electronic or mechanical, including photocopying, recording, or by any information storage and retrieval system, without permission in writing from the copyright owner.

Any people depicted in stock imagery provided by Thinkstock are models, and such images are being used for illustrative purposes only. Certain stock imagery © Thinkstock.

Print information available on the last page.

Rev. date: 04/20/2016

To order additional copies of this book, contact:
Xlibris
1-888-795-4274
www.Xlibris.com
Orders@Xlibris.com
738681

CYBER REALITY

The unprecedented realm of reality

By

Lou Giannelli

To my dear friend Cara...

... who taught me the power of faithfulness.

Disclaimer: The views expressed in this book represent the personal professional cyber articulations of this author, and are not necessarily the views of the Department of Defense or of the Department of the Air Force.

Table of Contents

Foreword..4
Preface...7
Introduction...9
Chapter 1. Ontologia and CYRE...12
Chapter 2. The Cyber Dimension.......................................21
Chapter 3. Cyber Identity...34
Chapter 4. Cyber DID...41
Chapter 5. Cyber UD...50
Chapter 6. The Legal Aspect of UD...................................69
Chapter 7. Global Cyber Identity.......................................75
Chapter 8. Cyber programming...85
Chapter 9. Abridged History of Computing.....................99
Chapter 10. The Scope of Cyber Reality..........................122
Chapter 11. The Perception of Reality.............................130
Chapter 12. CYRE Dislocation with Real Life................139
Chapter 13. CYRE and Big Data......................................145
Chapter 14. The IoE Impact..153
Chapter 15. Internet Governance.....................................172
Chapter 16. CYRE & Global Economy............................189
Chapter 17. Becoming a Cyber target..............................200
Chapter 18. Facing the Truth of CYRE...........................207
Chapter 19. A Taste of CYRE...214
Epilogue...223

Foreword

I'm not a cyber expert. I am, however, "cyber expert"-adjacent. Both as a career Air Force intelligence officer and now working in industry supporting the development of tools in the cyber realm, I've had to seek out cyber expertise in order to succeed. I live in this marketplace of ideas.

Claimed expertise in "cyber" is easy to find in the current marketplace. In contrast, expertise that is useful is challenging to find. Many experts will fall back primarily on their own (often limited) experience and beat you down with case study after case study from their personal life as if their experience transcends all situations. Others will mask their lack of comfort with the subject material by staying in the abstract and never tie into the real world or make concrete arguments. It is rare indeed to find someone who is willing to work in both abstract areas like well-defining terms and comfortable enough with the subject to drive that abstract concept back to practical suggestions. This usually also describes the rare breed of professionals who are willing to take controversial positions on debatable propositions, and in that breed, it's rarer still to find someone who is excited about teaching and mentoring the next generation.

One such leader in the US Air Force is my friend (and sometimes mentor) Lou Giannelli. His passion for cyber is infectious and you will always find him instigating thought, whether it's teaching his Advanced Cyber Course to new USAF cyber units, or simply one-on-one with someone who wants to understand basic cyber principles. In his writing, he attacks intellectual challenges, including the lack of common definitions, the nature of reality in the cyber realm, and the best way to characterize both vulnerability and advantage. It's not a book designed to be the final work on the subject, and often imparts a classical Socratic journey to help the readers organize their system of thought (as he says, he's less interesting in imparting answers than ensuring we're asking the right questions) in order to provide the next generation of cyber warriors advantage in the challenges that lay ahead.

His cyber expertise includes years as a net defender in technical workcenters, and is evident in the descriptions of the problem set. He's also an iconoclast, and is uninterested in aligning with a particular trend or buzzword. Lou isn't the guy you call to put the finishing graphics on your PowerPoint presentation to the Generals in charge of Cyber Command. He's the guy you call when you need someone to highlight the threats to the nations by asking boldly (even when those Generals are in the room) whether our current perception of cyber is even correct ... and the reason you want him in that room is he won't stop there, but will try to explain how an evolution of that understanding might drive both capability and vulnerability in the future.

Nobody has a lock on the truth, and as Lou tells it, the perception of reality in both life and in the cyber realm is both iterative and evolving. But what does help is every time you can pick up a set of questions that move the discussion forward, questions that you can incorporate into a set of ideas that challenge your current understanding and help you test your new one. For me, I can count on one hand the people who consistently have done that for me professionally, and Lou Giannelli is in that very small set. His work informs both the development of the next generation of cyber capability and the next generation of cyber leadership, and, I expect you will find, will challenge both your understanding of what you thought you knew about cyber, and make you reflect on what lies ahead.

Cyber Reality

Tim West, LtCol, USAF (Retired)
Senior Intelligence Consultant, PatchPlus Consulting.

Preface

There is an abundance of definitions identifying cyber as a new "domain", in addition to the other domains listed as air, land, sea, and space. However, the major flaw with the concept of adding cyber to the list of the other four domains, and furthermore, defining cyber as a domain, is the assumption that cyber shares the basic characteristics of those other four domains, as if cyber was an entity with a tangible existence predating humanity, and part of the natural infrastructure of planet Earth. Such an assumption is untenable.

Cyber is not a domain, but a dimension. Of the four traditional domains three are of earthly origin, defined by a specific biochemical environment conducive to the sustainment of earthly life forms. The fourth domain, space, is a sterile environment not compatible with life sustainment, but assimilated as a geopolitical environment accessible and available for human interaction.

Cyber doesn't share any of these characteristics, either generally or particularly, with those other four domains. Consequently, it is erroneous to call cyber a domain, since is neither a biochemical environment nor predates humanity as an earthly milieu. Those who insist in calling cyber a "domain" do so simply out of lack of proper information and understanding on the nature and essence of the cyber dimension. They resort to a frame of thought that it's compatible with their misconception of cyber, based on misinformed propositions. These individuals are familiar with the pre-existing domains such as land, sea, air, and then space, and they force cyber into the same category. Consequently, when the cyber dimension emerges into their collective consciousness, they simply follow the previous trend and existing model, and without proper analysis on the subject, and without consultation with those who could have offered guidance in their nomenclature, they defaulted into simply adding one more item to their familiar list; thus, the birth of the misnomer of what they incorrectly labeled cyber "domain".

The fact that an ample number of publications are issued under the label "official" with erroneous descriptions of cyber topics does not constitute a scientific or accurate description. Quantity in this case does not constitute quality. Does a concept popularly accepted constitute a *de facto* accurate definition of the entity represented in such concept? And how can we propose and accept a concept of a reality based on superficial, and often distorted, understanding of that reality? How do we dare to propose and publish cyber "concepts" without an in-depth understanding of the nature and essence of cyber? Very likely we feel incline to propose and accept "concepts" because we think that the definition and acceptance of a "concept" is to some extent the product of human cognition.[1] But how do we ascertain that our cognition of the reality we are attempting to conceptualize is based on an accurate and comprehensive cognition of such reality?

Cyber is not a domain, since is neither a biochemical environment nor predates humanity as an earthly milieu.

This book presents the reader with an opportunity to examine popular views concerning the cyber realm under a critical analysis, seeking for accurate concepts emanating from the essential nature of cyber, and its role in our modern society.

---

1 Barry Smith , Beyond Concepts: Ontology as Reality Representation, (Department of Philosophy, University at Buffalo, NY, USA)

Introduction

The concept of reality is an ontological conundrum, because it's intimately connected to existence, and every existence has both an objective and a subjective essence. The objective essence is connected to the collective consciousness, and the subjective one is connected to the individual awareness of self-existence. Is there a reality if there is no consciousness of existence? Is there a unified reality spanning both the macro and the micro dimension of existence? The discovery of quantum physics has taught us that our concept of reality at the macro level is quite different than the concept of reality at the sub-atomic level.

We have discovered through the millennia that human experience cannot be considered the final and conclusive factor to gauge our understanding of universal reality. In humanity, reality is always partial, tentative, and speculative. The early concept of the so-called four-elements conceptualization was eventually replaced by the more comprehensive table of elements, quite possibly still incomplete. Likewise, the early concept of a geocentric system was eventually replaced by the heliocentric model, and the articulation of matter as constituted by the atom model was eventually shattered by the discovery of a myriad of sub-atomic particles. Even our black-and-white attempt to discover matter interaction in our planet was demolished by the discovery of matter interaction as revealed by the research and discoveries in matter behavior at the quantum level. Yes, our concept of reality is still immature and unfinished, and in transition. Consequently, we must be prepared to submit our historical and collective concepts of reality to new evidence unveiled through our perennial quest for knowledge and understanding of our own and collective reality.

Cyber Reality

In the cyber dimension this quest takes on a sense of urgency, because of the dangers associated with accepting and adopting a pseudo-reality sustained by cyber misconceptions, an extremely dangerous predicament, considering the impact that the cyber dimension has on our collective and personal mode of existence.

Reality transcends collective and personal awareness; it even transcends theoretical and empirical knowledge, both collectively and personally. How many realities exist that transcend our awareness? Examples from our human history illustrate this condition beyond any doubt. How strong was the conviction of ancient people regarding their views of a flat earth? How strong is still this conviction among the modern followers of the flat earth model?[2]

So, let us ask ourselves: how do we perceive the cyber reality? How do we perceive the intricate and sometimes esoteric nature of the realm of binary code that constitutes the DNA-like building block of what exists and works in our cyber systems and networks? Do we really perceive the awesome power of the cyber code, not at the technical level necessarily, but at the ontological level?

Do we consider that cyber reality (CYRE)[3] is simply a matter of a technology placed at our disposal to do whatever comes to mind, free of all restrictions? If so, then we join the multitude of millions of people who think the cyber code exist solely for our convenience and entertainment, and thus, we perceive the cyber reality at an almost inconsequential level.

Or do we perceive that cyber reality is an unprecedented

---

[2] International Flat Earth Society, a modern organization founded in the late 19th century, and reestablished in 1956, currently still active and with a WWW presence.

[3] CYRE is an acronym coined by this author.

discovery only unveiled after placing the corresponding cyber code in deployment mode, interacting with systems and networks in ways that perhaps we didn't foresee during the code development phase? Then, we may belong to that minuscule population who realizes that cyber code has the potential of developing into an entity with characteristics transcending the limits envisioned by the developers.

These two views of cyber reality usually reside on diametrically opposite realms. The two groups rarely intersect their corresponding views of CYRE, even when these two groups may co-exist in their daily personal and professional lives. Thus, it is possible to visualize these two CYRE as two parallel cyber universes. Consequently, these two disparate views of CYRE creates a dichotomy in our real lives, where we see and conceptualize CYRe in two virtually antithetical realities.

Chapter 1. Ontologia and CYRE

The conceptualization and formulation of reality is in its purest form an ontological enterprise. We awake to the reality of being, and from within our Earthly point of view we formulate our relative and limited understanding of our personal being from our cosmological context. From this context we also formulate the conceptualization of origins, purposes and goals, and we strive to achieve social consensus in these areas at a local, regional, and global level.

The proper term "ontologia" (with its English transliteration ontology) is derived from the Greek ὤν, ὄντος (present participle of the verb εἰμί, to be) and λόγια (discourse, study). Thus, ontologia denotes the philosophical term dedicated to the study of being, existing, and cognition of reality, as conceptualized by the human mind. The term ontologia was used for the first time during the 17th century, on published studies dedicated to the science concerning the character of universal entities, the study of the being in itself, emphasized by philosophers as the fundamental truth above all others.[4]

Excursus
It is quite significant that the etymology of ontologia uses the peculiar present participle form of the Greek verb "to be". Greek grammar indicates that participles are verbal adjectives, and as such, instances of present participles indicates a synchronicity of action, portraying a simultaneousness. This unique condition of the Greek present participle requires the use of a temporal clause when translated into English. Thus, the etymology of ontologia necessitates the temporal clause "while" in order

---

4 Translated by the author from article on L'Enciclopedia Italiana, Treccani.it, http://www.treccani.it/enciclopedia/ontologia/

to convey the accurate meaning in English. Consequently, ontologia is properly understood when we focus on the study of a being while researching his or her own existing, simultaneously to the experiencing of being.[5] Based on this explanation, this author disagrees with the practice of misusing the term ontology when applied to the conceptualization of the hierarchical structure of a system not comprising human beings. There are countless organizations,[6] otherwise reputable, misusing the term ontology in their pursuit for an academic and scientific classification of components within a homogeneous system. For this endeavor the term taxonomy is adequately sufficient. Consequently, their misuse of the term ontology constitute a blatant disregard of the etymology and essence of the pristine terminology coined by the pioneers of ontological studies in the 17th century.

There is an intimate nexus between our personal and subjective ontologies and the cyber dimension. This is the result of the ineluctable correlation generated by our ontological conceptualization and our perception of reality in general, and our personal and developing conceptualization of CYRE. Our personal concept of CYRE is going to be formulated in close relationship to our personal concept of ontological identity. The awesome power of the cyber

---

5   The author holds a graduate degree with a concentration in Koine Greek

6   Leo Obrst, Penny Chase, Richard Markeloff, The MITRE Corporation, Developing an Ontology of the Cyber Security Domain, http://franz.com/agraph/cresources/white_papers/STIDS2012_T06_ObrstEtAl_CyberOntology.pdf; How Ontologies Can Help Build a Science of Cybersecurity, CERT Insider Threat Center, 03/12/2013, http://www.cert.org/blogs/insider-threat/post.cfm?EntryID=151; Journal of the National Institute of Information and Communications Technology, Vol. 58 Nos. 3/4 2011, http://cybex.nict.go.jp/publications/NICTjournal201203_cameraready.pdf; The First International Workshop on Security Ontologies and Taxonomies (SecOnT 2012), Prague, Czech Republic, http://www.allconferences.com/conferences/2012/20120222092106/

dimension will inevitably be subordinated to our concept of self. In plain language, this is what I think of myself, and this is how I see CYRE. My conceptualization of reality formulates the nexus explicating my conceptualization of CYRE.

Our knowledge of universal entities is predicated on our chronological milieu, since knowledge, for all human beings, is dependent on the process of discovery, and the subsequent interpretation of the discovered reality, which in turn becomes a conceptualization dependent on our chronological position in human history. What was accepted as knowledge and truth in a preceding historical period it may now be considered an erroneous concept. Our knowledge and conceptualization is forever located on a spiraling evolutionary path. After an initial concept is formulated, and a relative consensus is reached regarding that concept, the very same concept will be affected by a more in-depth discovery and understanding, leading to a revision of that concept, which will become redefined, or discarded.

Cyber Reality

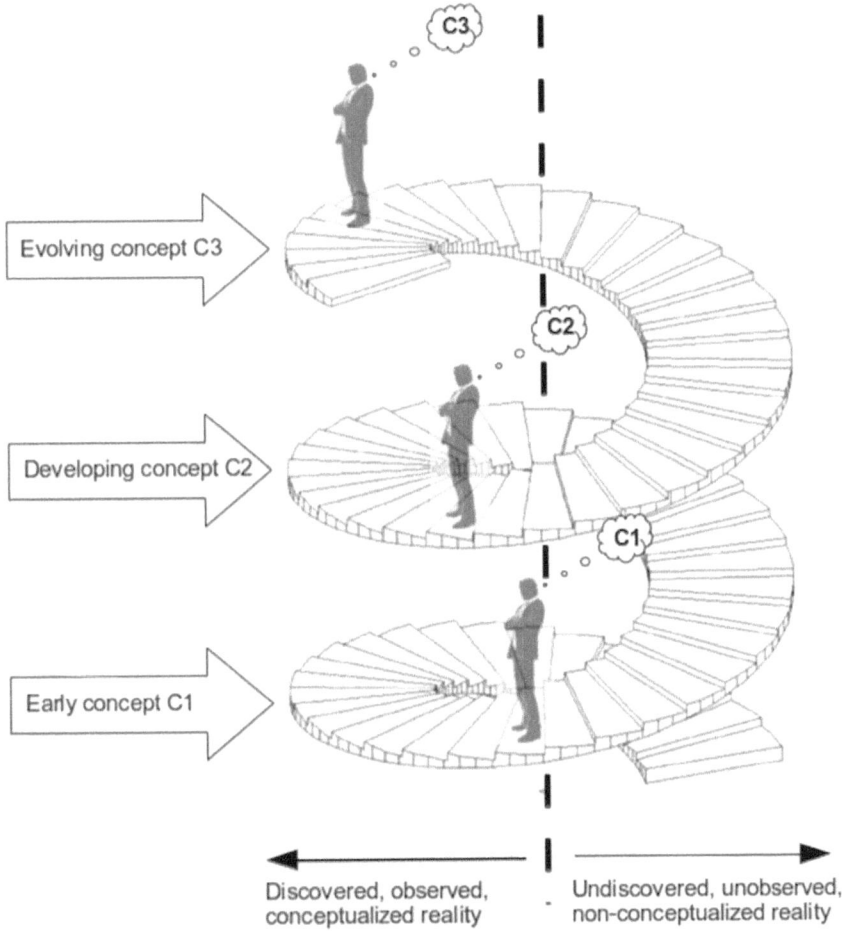

Graphical representation of spiraling conceptual progression

The spiraling conceptual progression illustrates the manner in which we perceive and conceptualize reality within our cosmological and contemporary reality, where concept "C" can reach "n" iterations, transcending the C1, C2, C3 series, since absolute knowledge is a human impossibility, due to the limitations of our temporary physical existence. Consequently, the actual equation of our conceptual progression is a function of the C1, C2, C3, Cn series, where "n" is any number of iterations, relative to how many times we review and enhance the process of discovery and conceptualization, leading to a series of spiraling levels of enlightenment based on the accessibility we may obtain to new data regarding the item under scrutiny.

The axis separating the two phases of the ontological experience illustrates the transitional levels of conceptualization. On the one hand (left) we have the transitional stages where we achieve a plane of enhanced understanding based on observation and analysis of what we have already conceptualized, and on the other (right) the ephemeral moments when we come momentarily in touch with a new enlightening opportunity to reassess and reanalyze our previously accepted concepts, leading to a new and higher level of cognition.

As the preceding graphic shows, we are limited to conceptualize only in past tense, based on the information already acquired. We may attempt a hypothetical conceptualization based on the purely intellectual exercise of extrapolation, but without ever reaching a purely ontological experience until we achieve an empirical level of interaction with the potential foreseen entities anticipated via extrapolation. A hypothetical abstraction, then, cannot be equated with an empirical conceptualization, because it remains in the milieu of pure human reasoning, that may correspond with a yet undiscovered reality, or it may simply be the result of fallacious reasoning. Can humans truly aspire to a completely objective cognition of reality? Can we truly say that what we have discovered, researched and formulated via conceptualization is the true representation of that allegedly unveiled reality? Have we completely unveiled it?

There's a parallel in the formulation of the concept of the atom as a misnomer, incorrectly portrayed as the basic unit of matter. The term "atom" is a misnomer improperly coined to indicate that which is considered indivisible, from the Greek ἄτομος, namely an entity that excludes any further division into smaller components. We now know differently, and this serves to illustrates the ephemeral condition of human conceptualization.

At the closing of the 19th century, and throughout the 20th century, scientists continue to borrow the ancient philosophical designator atom[7] in describing their observations in chemistry and physical experimentation, until the late 19th and early 20th centuries, when physicists discovered subatomic particles inside the so-called "atom" (indivisible) entity.

Human history has an abundant repository of concepts and realities affected by this evolutionary cognitive process. Once upon a time we though our planet was a flat surface, with edges we wouldn't dare to approach, less we fall into the unknown abyss beyond the borders. Until recently in our contemporary history, we accepted that the speed of light was the ultimate limit. Now, based on the findings by quantum physicists, we know that sub-atomic particles operate in a condition of instantaneity, as observed in multiple experiments involving the teleportation of entangled sub-atomic particles. The teleportation of entangled sub-atomic particles supersedes the speed of light, because there is no transference of matter in time. The teleportation phenomenon occurs instantaneously.

Quite likely the rather esoteric nature of cyber code, the essence of the cyber dimension, may contribute to the rather casual and trivial way in which 90% of the world population relates to the cyber dimension. For instance, cyber activity

---

[7] Coined by Democritus in approximately 450 BC, based on philosophical , not scientific, hypothetical conceptualization. See Democritus, Stanford Encyclopedia of Philosophy, Aug 25, 2010, http://plato.stanford.edu/entries/democritus/

is nonlinear, and this characteristic runs conterintuitively to the popular linear causality dominating the general conceptualization of a single cause and effect paradigm. As a result, general public may relate casually to the effects of cyber activity because in their personal ontologies there is a distorted conceptualization of CYRE. Thus, we ponder: is CYRE abstract or concrete? Since evolving knowledge exists only in the minds of human beings, and such knowledge is neither comprehensive nor absolute, then the realities assimilated through our perceptions are filtered, and sometimes distorted, through our presuppositions conditioning our personal ontologies. Thus, such perceived realities are not a representation of the true nature of such realities, but rather the representation of our own concepts. So again, we may ponder: Does an objective reality really exist beyond the boundaries on my ontological perception of reality? Are the effects of cyber code concrete and real in my own existence?

What are we saying? Isn't this a book on cyber reality? Doesn't this chapter sound more like philosophizing? Absolutely not! We are not engaging in a superfluous intellectual exercise attempting to elucidate an imaginary abstraction. We are simply setting the stage for the theme of this book. Only after we are able to formulate a cognitive conceptualization of our own being we can begin to conceptualize our understanding of the surrounding reality. When we consider a reality too abstract or ethereal we tend to limit that reality to the periphery of our interests, or we may acknowledge that particular reality at either a superficial or inconsequential level.

When we categorize a reality on an inconsequential level we may tend to view such reality only as a pseudo reality, as something that bears no consequences on our real existence, thus leading us to adopt an attitude that is excessively casual toward that said pseudo reality. We may think: "Well, it is out there, but it cannot bite me in real life; it's just a make-believe". Is this why we behave so casually toward the cyber dimension, and why we assume such trivial attitude toward cyber activity, thinking: "Well, cyber can only do things within the confines of a computer, but it can't really touch me on a personal level."

Our perception of reality is a strong factor that is shaping our behavior. Are we capable of conceptualizing coherent abstractions of a reality that remains unveiled to the self, or one that the self cannot rationally comprehend? Can we truly achieve as a global society a consensus on a given reality when confronted with the fact that realities are intrinsically attached to personal views, where the "my reality is not your reality" factor plays an active role in our perceptions and convictions? Can we truly share a consensual abstraction of CYRE? Obviously not, because we do not share the same perceptions and articulations of the different intricacies of CYRE.

The knowledge of CYRE is finite, and therefore potentially accessible to all human beings, but all human beings do not share the same degree of intellectual acuity on all the intricacies of CYRE. All human beings are potentially capable of experiencing snow, but all human beings do not actually share this experience. Consequently, articulating about the beauty, the joy, the risks and the threats associated with experiencing snow is beyond the possibility of achieving a consensual perception of such reality. So it is with CYRE: not everyone is equally prepared to experience the beauty, the joy, the risks and the threats of CYRE. Thus, our behaviors toward CYRE cover a vast gamut of diversities.

When we interact with the CYRE made available by the complex known as the Information and Communication Technology (ICT) in general, and the Internet in particular, can we truly distinguished between "real life" and "online life"? Can we really distinguished between "real life" and the global computerized system serving as the conveyance and operational universe where cyber code exists? Can we truly say that this universe is distinctly a "virtual world" dissociated from what we perceived as "real life"? Is an introduction via email less "real" than an introduction via a physical letter delivered by a messenger or by post office personnel?

This is not a book of answers. Rather, is a book challenging the reader to construct, articulate, and conceptualize a personal taxonomy of the universe where what we tend to perceive as real life and as cyber reality coexist in juxtaposition, or if we borrow from the lexicon of quantum physics, coexisting in superposition. With the arrival of cyber, life has become a little bit more challenging, a little bit more complicated, while a little bit more exciting and rewarding as well.

Chapter 2. The Cyber Dimension

The numerous voices that speak of cyber as the "other" or the "new" domain do so because of their incognizant condition on the subject matter. Cyber can never be categorized as a domain because it doesn't share the characteristics of a domain. The domains land, sea, and air are characterized by a unique life-sustaining biochemical composition, and when we add space, though not a life-sustaining but an inert environment, we group all four of them as domains because they all offer a milieu for human affairs. For the sake of geopolitical and military boundaries we define a traditional domain as the environment that is either under the control of a particular nation, or existing as an international heritage. Cyber transcends all this definitions and characteristics. It is highly erroneous and disingenuous to group cyber with these traditional domains.

Many accepted definitions fail in recognizing the three-dimensional characteristics of a domain, by defining it as an "area" of activity under the jurisdiction and control of a given entity. Recently, and only after recognizing the importance of cyber activities in human affairs, there has been a concerted effort to add to the traditional domains (land, sea, air, and space) a fifth member, namely, cyber. This arrangement. however, is both erroneous and incongruent with the unprecedented essence of cyber. Cyber is not a natural Earthly environment preceding humankind. Cyber is the sphere of activity where cyber code is generated, deployed and executed, and this as part of a human invention in the late 20th century.

Myriads of spokespersons, with a very superficial scientific and technical knowledge of the essence of cyber, simply borrow lexicons from known traditional warfare in order to articulate their thoughts regarding cyber activity. In doing so they negate and disregard the uniquely unprecedented nature of cyber, that as such, it demands an equally unique and new taxonomy. CYRE is a dimension encompassing and transcending all the four natural domains of human affairs.

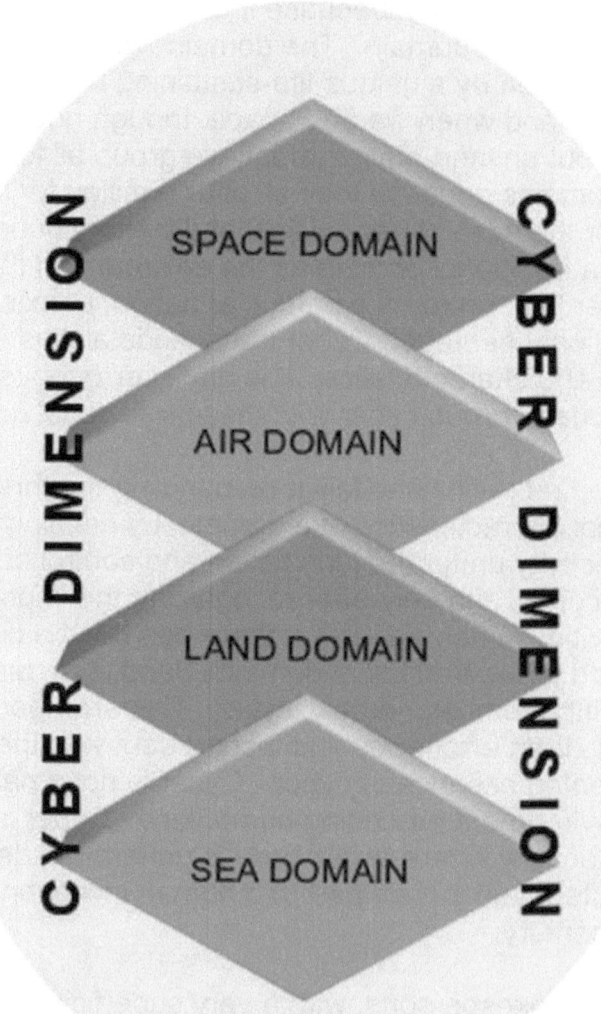

The transcending cyber dimension

There are very few published articles using the term "dimension" when dealing with the realm of cyber activity. But even those few use, and abuse, the term dimension as a synonym for the term domain. Consequently, the conceptualization of cyber as the sphere of activity substantially different from the other traditional domains referenced in countless warfare taxonomies remains a problem, despite of the alternate terminologies erroneously applied to CYRE.

Cyber is not another domain to be added to the previous list. Cyber is a completely different and unprecedented dimension, a unique sphere of activity that cannot be inferred from legacy concepts and lexicons used in traditional warfare. Just as quantum physics requires the use of a new vocabulary and conceptualization to properly understand it and research it, so does cyber. It is a new science, a new technology, a new reality, with rules and behaviors without antecedents, and cannot be treated as a legacy system with parallels to traditional concepts associated with principles of traditional warfare.

From a plethora of published cyber articles portraying the disingenuous categorization of cyber as the "other domain" we can see how widespread this misconception literally is. Perhaps nothing serves better the purpose of illustrating this misrepresentation than looking into published material speaking for the international organization NATO. In 2011 their official web site posted an article addressing the issues associated with the mislabeled cyber domain or cyber dimension.[8] The spokesperson used the terms "domain" and "dimension" interchangeably throughout the article.

---

8 Olaf Theiler, New threats: the cyber-dimension, NATO Review Magazine, 2011, http://www.nato.int/docu/review/2011/11-september/Cyber-Threads/EN/index.htm

Cyber Reality

The problem of misrepresentation of cyber as a domain goes beyond the choice of poor and inappropriate terminology. Every one of the four traditional domains, i.e., land, sea, air, and space constitutes a different area of operations with a peculiar set of strategies and equipment required by the physical characteristics of each environment. The training and deployment of personnel designated to operate in any one of these distinctly separated environments requires specialized knowledge, tactics and techniques, unique to each one of them.

The cyber dimension, on the other hand, transcends all these other natural environments, and transcends the natural characteristics and limitations of each one of them as well. The cyber dimension, when properly, scientifically and technically conceptualized and assimilated, does not share the characteristics of the natural domains. Yet, the cyber dimension, as a man-made sphere of activity, can perform beyond the natural boundaries of the traditional domains in the taxonomy of traditional warfare. The cyber dimension does not belong to any of these domains, and transcends all of them, while at the same time being capable of utilizing any infrastructure created for operations designed for each one of the four natural domains.

Cyber is not just another domain, simply because is not just a natural Earthly environment preceding humankind. Cyber is the realm of activity where cyber code is generated, deployed and executed, and this as part of a human invention in the late 20th century.

Why is it that people adopt such an impervious and resistant attitude against the uniqueness of cyber? The obstacle is definitely not intellectual capacity, but rather willingness. The newness of cyber upsets our comfort zone, and sends our preconceived notions and categorizations into an unwelcome upheaval. We don't feel comfortable accepting the newness of cyber, so we try to force it into one of our well-established models of warfare. We think: "Hey, we have four domains in our models, and here it comes this new thing called cyber... Oh, it easy, let's just hang another peg on the wall, alongside the four other ones, and now we have a fifth one to hang from the new additional peg. We will fit it, even if we have to force it, into the same model than the others. Problem solved."

If we follow this rationale in dealing with realities that are different than other traditional realities, what are we going to do with quantum physics, and all the new technologies derived from quantum physics? Are we simply going to deny that subatomic particles behave in different ways than matter in our macro cosmos, simply because such behavior doesn't fit our traditional models? Are we going to slam the door of our tree house to the new realities we encounter? CYRE is not a new thing; it's been with us for the last 60 years. That's pretty much a life time! How much more time do we really need to understand this new technology and accept it for what it is, instead of trying to force it into our traditional models?

And when in the very near future our gifted researchers bestow upon us the gift of a productive manner to harnessing the awesome computational power of quantum computing, are we going to be thrown once again into the backward-looking mindset of those who resist the peculiar milieu of CYRE? Are we going to be bombarded by the

Cyber Reality

empty rhetoric of those who try to force the new into old molds, compelled by their desperate attempt to maintain their comfort zone, alien to new technologies, requiring the corresponding new conceptualization and lexicon? Are they going to encounter the arrival of quantum computing by placing a sixth peg on the wall, and force it to fit the old models?

Quantum computing (QC) will occupy a higher level of cognizance into the spiraling conceptualization of reality. QC no longer relies on the binary system that gave birth to classical computing. QC operates on quantum bits (qubits), an exponential departure from the traditional bits, the unit of computational calculation in classical computing. If as a global community we are still holding to misconceptions regarding CYRE, after some 60 years of interacting with its reality, are we going to spend another 60 years exposed to misconceptions upon the arrival of the QC reality?

Quantum Information Science & Technology (QIST) is here to stay, and it brings with it a vast array of disciplines, including, but not limited to, quantum cryptography, quantum key distribution (QKD), quantum computing, quantum algorithm, quantum coding, quantum entanglement, quantum teleportation, quantum complexity theory, etc. Are we also going to insist on forcing these new disciplines and their new required lexicon into our old models, so as to maintain our cozy comfort level and not upset our traditional concepts of reality? Reality is the continuum of our interaction with the environment of our universe, as we progressively discern, and conceptualized our new discoveries and experiences. As we exist, we think, and we become aware of our own identity interacting with the discovered reality. And then, as we think, we conceptualize, and as we conceptualized, we develop a sense of reality that is not static, but dynamic. We move forward, we adapt to new discovered realities, and we begin the ascension to the new spiral level in our existence (see chapter 1). Once we reach the new phase of existence into our new reality level, we begin the process again: we exist, we think...

When searching on current writings on these topics is apparent that most authors have chosen a static position in the spiraling conceptualization of reality. It is evident that there is no forward momentum toward the acceptance of the different reality of cyber. There is a myriad of other authors endowed with only a superficial and peripheral interaction with adversarial cyber activity, but they feel entitled to act as spokespersons, and proceed to pontificate on pseudo analysis of cyber conflict cases or trends.

One such author states that "the most dangerous actors in the cyber-domain are still nation-states."[9] This statement reflects a shallow understanding of what constitutes cyber prowess. The most dangerous cyber actor is the one with the most profound knowledge on a particular cyber attack methodology. The few of us who have encountered the cyber adversary in the course of a cyber incident know that the key factor in cyber victory resides on the knowledge and skills of the human being directing and strategizing the adversarial cyber action. Cyber prowess is an individual quality, not a nation-state quality. Nation-states may contribute to the organization, deployment and support of cyber operators, but the success of the adversarial action rests on the shoulders of every individual cyber operator, whether in the offensive or defensive role.

In the cyber dimension there are primarily two categories of practitioners: observers, and operators. If you have an in-depth empirical knowledge of binary language, networking topologies, and networked and networking protocols,[10] then

---

9   Ibid
10 Networked protocols are used by the many applications generating networked data flow, while networking protocols are used by the

you are an operator. If you don't, you are simply an observer. Approximately 80 percent of practitioners belong to the latter category. Unfortunately, these observers are also the most loquacious.

Access to knowledge of CYRE is readily available to anyone with the commitment to learn. CYRE is neither a secret knowledge, nor part of an exclusive elite. As a matter of fact, CYRE knowledge is literally available on the palm of our hands. Just select any proper and accurate cyber term, pick up your smart phone, and enter the selected term into the Google search engine. You will receive a vast number of answers elucidating the selected term. Anyone can learn, without cost, about CYRE. Of course, the free availability of CYRE knowledge can be defeated by an apathetic and unwilling attitude toward learning, and against that there is no cure...

The key factor in cyber victory resides on the knowledge and skills of the human being directing and strategizing the adversarial cyber action. Cyber prowess is an individual quality, not a nation-state quality.

Another misguided common opinion states that "the cyber warfare domain ... is asymmetrical, ... and all the advantages are initially on the attacker's side. "[11] Yes, there it is, asymmetry, the very favorite word of those with an insufficient knowledge on cyber activity, who are still clinging to their legacy terminology, irrelevant in the cyber sphere of activity! Of course they prefer the term "asymmetrical", coined and adopted from unconventional war conflicts resulting in the victory of the weak opponent against a

---
network infrastructure devices responsible for regulating the flow of data.
11 Ibid

stronger opponent. This legacy war terminology is irrelevant in the cyber dimension, and its use, or better yet, misuse, it's fallacious.[12]

Only a casual observer would say that the advantages reside with the adversary in the context of a cyber attack. This may be a correct assessment only on a network environment configured exclusively with automated detection tools, the only counteractive resource sought by those who don't know how to design and implement an intelligent, proactive and effective cyber defense strategy.

A cyber encounter is predicated on the rules regulating the use of binary code and networked and networking protocols. There is no magic wand giving an *a priori* advantage to either side, the offensive or the defensive. Either side is equally equipped to detect the activities, and even the intentions, of the other side, if, and only if, either side is acting under the advice and strategies generated by qualified and certified cyber subject matter experts (SMEs). Obviously the cyber defensive scenarios available to the NATO writer clearly indicate an appalling lack of the required cyber SMEs, and in the absence of them he can only envision and generalize his gloomy and defeatist outlook.

> ***Those who remain in the periphery of the cognitive envelope of cyber understanding are easily overwhelmed by the magnitude and frequency of cyber attacks, and find solace in dreaming of powerful automated detection systems that will protect them from the attacks they do not understand. Those of us who work inside the cognitive envelope of cyber understanding do not despair at the sight of the avalanche of cyber exploits. Instead, we***

---

12 See Giannelli, chapter 6

> *contribute to the design and building of tailored cyber defenses, based on a layered cyber security posture that integrates both automated detection systems and cyber SMEs, working in synergy with cyber security strategies designed by certified and experienced cyber warriors, and deployed and implemented under the direction of cyber SMEs. The best of both worlds is the paradigm of fusing the work of automated detection systems and the analytical skills of cyber SMEs, in a unified task where the human being, not the automated system, has the last word in cyber defense.*[13]

Finally, we are also presented with the incongruous advice and erroneous view that "there is virtually no effective deterrence in cyber warfare..." Of course there is no deterrence for those holding the view that cyber defense is a passive and reactive cyber defense! This model, hardly deserving the designation as defense at all, is based on a construct depending exclusively on automated systems, and a contingent of unqualified and inexperienced individuals dedicated to simply report on the number of incidents occurring on the network environment affected by the adversarial cyber activity. The paradigm cited in the previous quote leads to a proactive cyber defense that allows not only for a successful deterrence, but also for a very successful mitigation of cyber attacks. When the brains of well qualified and certified cyber SMEs are at work in defending a network environment, applying an analytical and proactive cyber defense model, the results include not only effective deterrence, but successful reduction on cyber attacks.[14]

---

[13] Giannelli, chapter 9
[14] The author of this book has documented statistical data showing a reduction of attacks over a period of five years, from an average of

The articulation from the cited NATO writer[15] is a typical example of the misconceptions circulating among international circles. Regretfully, given the influential position these individuals hold in the cyber arena schema, their opinions are broadcast, and even worse, accepted, as authoritative views. How, then, do we build a more coherent and insightful view into CYRE? By seeking and articulating unambiguous definitions addressing the proper conceptualization of CYRE based on the very essence of CYRE.

A very insightful paper[16] addresses the problem of ambiguity in defining the multifaceted aspects of CYRE, and the associated costs and risks brought by these ambiguities. At the global level, governments produce a deluge of documents addressing the importance of ICT, as the basis for the management and defense of the critical infrastructures sustaining modern societies, subject to the effects of cyber malfunctions or adversarial cyber attacks. However, the positive impact of these documents is diminished by the presence of unclear definitions and lack of harmonization on cyber-related concepts, resulting on negative impacts on the implementation of corrective and defensive cyber security measures.

The authors of the Istituto Affari Internazionali (IAI) paper highlight that their research on recent documents on cyber matters, published by the US and the EU, indicates a widespread problem regarding the absence of accurate and specific cyber definitions. They state unequivocally that "an

---

175 incidents per year, to 5 incidents per year.
15 Theiler, who also shows disregard for protocols safeguarding restricted material that should not be posted on public web sites.
16 Federica Di Camillo and Valérie Miranda, Ambiguous Definitions in the Cyber Domain: Costs, Risks and the Way Forward, Istituto Affari Internazionali (IAI), September 2011, http://www.iai.it/pdf/DocIAI/iaiwp1126.pdf

Cyber Reality

explicit and exhaustive definition of the term cyber-security is never once provided, [but] used predominantly as a blanket term."[17] Given this lack of specificity, it negates the possibility of establishing the exact scope and boundaries of cyber-security, and clearly distinguish it from other frequently used derivative hyphenated cyber terms, referenced in a variety of contexts and ambiguous manners.

The corollary of this ambiguity is the introduction of a misleading factor when assessing risks. Cyber ambiguity may result in the allocation of resources to cyber events with only improbable occurrences, while neglecting cyber events with an impending and greater impact on the well-being of the population.

Let us summarize the items discussed in this chapter. Cyber is not "another domain" or "the fifth domain" as referenced on an ever growing mass of published opinions. Cyber is on a category of its own, because is without antecedent and precedent, and is not part of the Early environment. Cyber is the man-made dimension, the sphere of activity of binary code circulating and interacting in the global Internet environment, and in quasi-isolated cyber systems.

Cyber ambiguity may result in the allocation of resources to cyber events with only improbable occurrences, while neglecting cyber events with an impending and greater impact on the well-being of the population.

Interaction with the cyber dimension requires a new mindset, a new skill set, a new lexicon, a new approach, a new willingness and attitude focused on learning CYRE. In sum, it requires the recognition as a new reality; the realm of CYRE. Those who resist the newness of CYRE are

---

17 Ibid

condemned to misinterpret and miscategorize it, and those who self-nominate themselves as cyber spokesperson, while lacking the proper specialized qualifications required by CYRE, will condemn their audiences to perpetual misrepresentations of CYRE, with long-lasting adverse effects.

Chapter 3. Cyber Identity

We seem to be traveling on an upward spiraling path in our collective quest for establishing a cyber presence as members of the cyber society. In many cases, this quest for a cyber identity may adopt the form of multiple cyber presences, including multiple email accounts, multiple social networks (SN) accounts, and smart phones to extend our cyber presence beyond the confines of our personal residence. Is this perhaps a subconscious quest to achieve ubiquity, and transcends the limitations of our localized and ephemeral personal reality? Are we perhaps seeing a glance of the Icarus complex,[18] characterized by complacency and excessive pride and over-confidence? When we are enabled with technology we may also receive a boost to our innate pride. The resulting impact may translate into an imbalance between one's desire for notoriety, success, and achievement, and one's ability to actually achieve such goals.

Airports have become a smörgåsbord of cyber identities, with transient individuals carrying and using several devices allowing them to stay connected with the cyber dimension. It is quite common to see on-transit travelers carrying a laptop, a tablet, and two or three smart phones. These devices may provide a degree of required connectivity for the traveler, but there is also another dimension that requires exploration and justification. Do we really need all those devices to carry on with our business in an uninterrupted manner? And if so, can we truly carry on with our business in a public place, where the environment denies us of all possibilities for privacy? The other alternative is the consideration of a

---

18 The mythological event of "Ἴκαρος" (Íkaros), the son of Daedalus. Icarus received from his father a set of man-made wings to escape Crete, along with the warning not to fly to close to the Sun. He disregarded this advice and paid with his life.

psychological compulsion to carry all those devices in order to proclaim and project our personal and professional importance, thus separating us from the collective anonymity pervading all crowded and public places. And if indeed we are that important, how do we explain that we are still traveling using public transportation, instead of using the convenience and expediency of a private airplane? These considerations do indeed introduce the question of the psychological aspect of the cyber dimension.

There are sufficient reasons to advance the working hypothesis that cyber identities are transcending their original utilitarian purposes, and they have become an outlet to proclaim our existence to the entire online world, reaching levels of exposure and intensity that betray the limits of personal privacy. More and more individuals are using their cyber identity to unveil layers and layers of personal information. Regretfully, this compulsory appetite for disclosure of personal information exceeds the boundaries of personal safety, by revealing aspects of our identity that eventually will expose us to cyber attacks designed to steal our identity, the ultimate danger in the cyber dimension. This dangerous compulsion to self-disclosure may also indicate a pathological side of the cyber identity reality.

The way in which we conceptualize our cyber identity is a driving factor in the way we behave in the cyber dimension. Do we hold the misconception of envisioning our activity in the cyber dimension as if we are bestowed with a cloak of invisibility, and no one can see us and determine who we are? Cyber operators can easily determine not only the identity of a cyber observer,[19] but the location where they are geographically conducting their cyber activities. A very good friend of mine spent some time overseas, and when she returned to the United States she was vary surprised I was able to tell her I already knew of her return, even though she

---

19 See chapter 2 above for a review of these two categories.

Cyber Reality

had not yet told me she was back. How did I do that? A simple analysis of the SMTP headers[20] on her last email, showed indicators that this last email originated on mail servers located in the United States, not overseas. Thus, I knew she was emailing me from a mail server in the US.

There is no cloak of invisibility in the cyber dimension, and even more, there is no complete privacy in the content of the information we decide to entrust to the Internet infrastructure. Do you have a "free" web-based email account? Or do you have email service provided by your particular ISP? Do you think that you have absolute privacy on your cyber identity and cyber activity? You do not.

There is no absolute privacy for your cyber identity or cyber activity. You can only enjoy a degree of privacy, and this degree is determined by a series of factors, including but not limited to, your behavior in the cyber dimension, and the configuration settings on the portion of the Internet infrastructure owned by your ISP. Let's consider a few popular cyber activities and their impact on your cyber identity and privacy.

You regularly email your family members, friends, acquaintances, and professional contacts. So, let's start with your email account: Are you the kind of person who prefers to create an account using your full legal name, or part of it? Are you the kind of person who prefers a password that it's easy to remember because is connected to your personal life (birth date, pet name, hobby, profession)? Do you realize how easily your cyber identity can be compromised by maintaining these practices? I have helped quite a number of persons who fell victims of identity theft, and in 95% of the cases their email accounts contained either their full name or

---

20 The SMTP is the cyber protocol facilitating the transmission of email. The SMTP headers are invisible to a cyber observer, but available to a cyber operator.

part of it, and their passwords were extremely obvious. Furthermore, if you openly broadcast information about personal identifiers as the ones listed above, then guessing the password of your email account is trivial, and any one can potentially impersonate you in the cyber dimension. In chapter 10 we discuss the issues regarding PII, an unavoidable aspect of our cyber identity.

The proliferation of your cyber identity in social and professional networks leads to an exponential increase in personal identity impersonation. If a malicious individual wishes to use your name and influence on a particular social or professional online group, all it takes to impersonate you is to reference your name, harvested from the enormous amount of online correspondence with a large distribution list. All it takes is a simple statement such as: "Paul, Susie tells me that ..." and the malicious impersonator has a high probability of eliciting information from Paul, simply because of the trust relationship falsely created by referencing Susie's name, who is part of the social or professional circle infiltrated by the impersonator.

The principles of social engineering, masterfully executed by renown practitioners such as Kevin Mitnick, can be easily applied to cyber identity theft, by capitalizing on the enormous amount of personal information Internet users disclose about themselves through sites designed to broadcast personal interests, such as Pinterest. The practice of social engineering is the procedure of exercising deceptive intellectual and psychological influence on people with the goal of eliciting confidential information from them. At the closing of the 1990s decade, Mitnick earned the number 1 position in the list of the most-wanted computer criminals in the United States.[21] After completing his incarceration period, Mitnick became a security consultant, and occasionally wears a black T-shirt reading: "I'm not a

---

21 http://en.wikipedia.org/wiki/Kevin_Mitnick

hacker. I'm a security professional." This is not the only case. There are numerous individuals who have decided to explore the legitimate side of cyber business, after experiencing the opposite side.

Social engineering remains as a viable technique for eliciting information leading to gaining access to systems with restricted access. Do you want to know about a person's interests? Visit Pinterest, locate the person in question, and you will see a complete visual display of the interests of that person, on virtual boards designed to show and share collections of ideas, preferences, trips, organized events and projects. Even better, you can get the names of the people joining and sharing interests with that person, and now you have instant access to a circle of friends and their shared personal preferences. Apply social engineering techniques, and you can gain access to their cyber identities.

There is no cloak of invisibility in the cyber dimension, and even more, there is no complete privacy in the content of the information we decide to entrust to the Internet infrastructure.

The more you disclose about yourself online, the more vulnerable you become. After all, the data you publish online is no longer yours. It belongs to the cyber dimension, to the owners of the cyber systems where you placed your information, and you cannot delete it! Once we write data to a cyber system, that data is replicated to other systems, thus negating the possibility of eradicating the posted data. Exercise caution, proceed with wisdom, think twice, even better, thrice, before you post data revealing details about yourself! What you post online is leaving bread crumbs leading directly to your door.

The illusion, the dangerous illusion, is to consider the cyber dimension as a world of make-believe. This misconception

Cyber Reality

then leads us to a distorted sense of reality. We come to believe that CYRE is not real, and that entering the cyber dimension places us in an anonymous dimension, and because of this misconstrued sense of anonymity be behave differently online. We create the illusion that we become invisible in the cyber dimension, and that our actions will have no repercussion in our real physical lives. This illusion of invisibility and anonymity then leads us into thinking that when we act online we are shielded from being identified. Thus, since we are acting as if we possess two different identities, two different personalities, one in "real life" and the other "online", we are actually transitioning into a dualistic identity; we become DID-affected entities, with two or more personality states.[22]

Furthermore, since the cyber dimension allows us to communicate with anyone around the world, we create another layer of false assurance, erroneously thinking that if we establish a cyber line of communication with our antipodean,[23] we will be protected by a buffer of great distance between us. In the cyber dimension distances are blurred, and we all live in very close proximity. Malicious code is not impeded by distances and geo-political boundaries. In the cyber dimension the world literally becomes a village. Distances are no longer a buffer between two or more communicating parties, and we are misled by a false sense of security of what we assume to be the distance between these parties. Our actions online will have direct consequences in our real life. Any online conduct leading to a legal offense, and its associated cyber identity, will become identifiable and liable before the law. Just consider the case of the young man who threatened a high school via an Internet message, and was sent to prison

---

22 Psychology Today, Dissociative Identity Disorder, http://www.psychologytoday.com/conditions/dissociative-identity-disorder-multiple-personality-disorder.
23 The person leaving diametrically opposite in the world.

for his offense.[24]

While recently attending and presenting at a cyber conference in London, I was invited to join a professional panel to discuss aspects of cyber identity. One of the panel members voiced his opinion that there seem to be indicators pointing to the presence of a cyber "schizophrenia"[25] among a significant percentage of the cyber population, represented by the fact that cyber users dissociate cyber from real life, behaving as is cyber existence and real existence were two separate dimensions. Cyber existence is a different expression of real life existence, manifesting itself in a digital environment, but intimately connected with our real life existence. To dissociate these two environments as if they were separate is a pathological and dangerously misleading perception.

Is this author simply hyperbolizing this dissociation problem by using the term DID? As of 2014, there are controversies surrounding the diagnosis of dissociative identity disorder among psychiatrists.[26] So, am I really hyperbolizing?

---

24 The Chicago Tribune, Man Who Threatened Columbine Sentenced, April 29, 2000, http://articles.chicagotribune.com/2000-04-29/news/0004290102_1_michael-ian-campbell-columbine-student-columbine-tapes.
25 The reference to schizophrenia was not done in a scientific context, but simply as a hyperbole for the phenomenon affecting cyber users. The proper clinical term DID is used elsewhere in this book.
26 Paulette Marie Gillig, Dissociative Identity Disorder. A Controversial Diagnosis, National Center for Biotechnology Information, http://www.ncbi.nlm.nih.gov/pmc/articles/PMC2719457/

Chapter 4. Cyber DID

As of the moment[27] of this writing there are no diagnosed cases of cyber dissociative identity disorder (DID), but it won't be a surprise when an entrepreneurial therapist will resort to this diagnosis in order to condone someone's cyber irresponsibility and disguise it as a cyber DID case, since a significant percentage of the world population behaves as DID-affected individuals when traversing the cyber dimension. I'm aware that by the very act of writing this sarcastic comment I may also be planting the seed for the occurrence of the very event I'm dreading, but denouncing cyber irresponsibility outweighs the risk.

This author prefers the use of the term cyber DID not as a hyperbole, but rather as a realistic depiction of the problem affecting many cyber users. The malady DID is the psychological disorder that affects its victims with dissociation in rendering what is real and not real. Cyber DID, on the other hand, is not a psychological dysfunction, but a behavioral choice. Cyber users engage in a behavioral pattern where they refuse to accept that cyber actions have consequences in real life. During the recent London cyber conference[28] I cited in the previous chapter there was a unanimous consensus in assessing that one of the main problems impacting cyber security efforts is the behavior of users at all levels, failing to correlate the connection between their online actions and the corresponding consequences in real life. Since the discussion of this topic took place during a professional meeting of cyber experts, the issue on cyber DID is no longer just the opinion of this author, but one that merits the attention of cyber professionals during international meetings.

---

27 June 23, 2012
28 "Cyber Defence 2012", 18 – 19 June, 2012

Cyber Reality

Users entering the cyber dimension tend to adopt the attitude of embodying an alternate cyber persona that remains immune to consequences in the real world. The cyber dimension is an extension of the real concrete world, and every activity in the cyber arena can be associated with the concrete real identity of the user in the real world. The cyber dimension is simply another dimension of existence, but it is neither imaginary nor fictional. It is simply an extension of our daily life where we can act and interact through the cyber dimension infrastructure. However, the port of entry into this cyber dimension is firmly rooted into the real world. To enter this new dimension we create a single or multiple cyber identities that become part of our identification in the cyber dimension. These identities are always associated with our physical persona.

Let examine briefly the infamous case of social networks sites (SNS). We are given the opportunity of connecting ourselves with a myriad of other entities in the global Internet village. We create a cyber persona for ourselves in the SNS of our choice, and then we proceed to travel, greet, and befriend other real or pseudo-entities. Since this cyber exchange in the SNS is usually limited to remote cyber interaction, there is very scarce evidence for us to assess the reality and trustworthiness of the remote cyber entity. It's actually very enticing to immerse ourselves into the idea of creating an alternate pseudo identity for ourselves, and traverse the cyber medium contacting, connecting and exchanging with other cyber entities that may also represent an alter ego.[29]

The proliferation of web-based SNS constitutes the enabling

---

[29] Term coined by Marcus Tullius Cicero, one of the greatest minds of ancient Rome, who used the term to designate "a second self, a trusted friend". See Cicero's Letters to Attico, D. R. Shackleton Bailey, Vol II (Books III-IV, Cambridge University Press 1965), 67

factor for the social connectivity explosion that has become the proverbial double-edge sword dilemma; a capability that both help and hurt the users. SNS has facilitated the hunting ground for dishonest users and cyber criminals, providing them with one of the most effective means to target and exploit anyone around the global cyber village. Malicious SNS predators have now the opportunity to search, target, and exploit their victims, by either writing malware designed to infiltrate their victims' cyber systems, or manipulate and mislead their victims through the subterfuge of the so-called social engineering techniques. Such techniques allow malicious actors to obtain information facilitating the exploitation of social and/or professional connections. There is a concise and simple advisory guide offering advice on becoming aware and cautious regarding the inherent dangers in using SNS.[30] Responsible cyber observers would do well in adhering to this guide.

The immense attraction of using SNS has also contributed to the rendering of SNS as the dissemination agent for spyware and malware, thus enabling cyber criminals with an enormous installed base for the propagation of botnets. Perhaps the single most damaging negative effect of SNS is the facilitation of a global platform for data leaks, encompassing both personal and professional data. Usually the former becomes a liability for personal credibility and embarrassment, while the latter is persistently becoming the source and catalytic for diplomatic and national security issues.

The most pernicious effect of SNS is the irreparable damage on the trust factor that is required for data exchange. The many threats enabled by SNS are demolishing the concept of trust, considering that the true identity of a SNS users is not a guaranteed fact. This is the result of the numerous

---

30 U.S. Department of Justice, Federal Bureau of Investigation, internet-social-networking-risks.pdf

cyber threats affecting SNS,[31] contributing to the creation and sustainment of a great deal of impersonations that go undetected by the SNS users being targeted by cyber criminals and adversarial agents.

> ***Cyber DID is a behavioral choice, occurring among cyber users choosing to engage in a behavioral pattern where they refuse to accept that cyber actions have consequences in real life.***

What does the SNS phenomenon have to do with CYRE and the cyber DID phenomenon modifying the behavior of so many online users? In a concise manner, the SNS phenomenon has simply introduced a rather radical change in our daily reality, and thus, a significant change to CYRE and our ethical principles. Our professional lives are no longer protected by a degree of separation from our personal life styles. There is a great percentage of professionals that continue exchanging professional data after hours at the office, simply because the interlocutors can now gather at the SNS playground where they can mix personal and professional issues. And with this virtual playground atmosphere comes the relaxing of the rules that, otherwise, used to protect the professional data we used to exchanged in a more controlled environment provided by the working place. Ethical behaviors are becoming fragmented and diluted, and what we considered unethical in our personal and professional lives becomes a feasible possibility in an ethical model ruled by relativity. We dissociate our real life identity from our online identity, and ecco qua,[32] we are now cyber DID cases!

---

[31] http://www.networkworld.com/news/2010/071210-social-network-threats.html contains an insightful article on these threats.
[32] The Italian interjection: That's it!

SNS have provided the forum for disclosure of sensitive information by government personnel displaying very poor judgment. An Israeli soldier posted on Facebook details of an impending West Bank operation. His action resulted in the abortion of the mission.[33]

In a well publicized case of SNS-entity impersonation, a significant number of people responded to a fictitious SNS persona created as part of an experiment to expose the risks of careless SNS contacts. The Robin Sage Experiment, a 28-day research project ending in January 2010, resulted in the disclosure of personal and enterprise information revealed by numerous SNS users contacting the fictitious Robin Sage.[34] In one particular case, an American Army ranger exposed information about his location in Afghanistan by uploading to Robin pictures containing geo-location data from his camera, in violation of operational security (OPSEC) protocols.[35] If Robin would have been a terrorist agent, this information could have been exploited with fatal impact. SNS users are most of the time so engrossed in disseminating information that they fail in exercising sound judgment and discretion.

We may question the actions of this Army soldier by asking: what was he thinking when he did that? Perhaps is relevant to offer the alternative question: WHO was thinking when he did that? Was the soldier meeting face-to-face with a female and revealing restricted information to her? Or was this soldier's cyber alter ego, his second self, the one this soldier becomes when he envisions himself acting in the cyber dimension, an entity dissociated from his "real self" because he's acting now as a cyber DID case? Is this an issue

---

33 http://www.foxnews.com/tech/2010/03/03/israeli-raid-called-facebook-slip/
34 Thomas Ryan, "Getting in Bed with Robin Sage", (presenter at Black Hat USA 2010).
35 http://www.darkreading.com/privacy/robin-sage-profile-duped-military-intell/225702468

affecting only low ranking soldiers? If you are inclined to think this is the case, read the next paragraph...

The fictitious Robin received contact requests from personnel in the Joint Chiefs of Staff, the NSA, an intelligence director for the US Marines, a chief of staff for the US House of Representatives, and several Pentagon and DoD employees. From the clear defense contractor (CDC) pool Robin also attracted personnel from Lockheed Martin, Northrop Grumman, and Booz Allen Hamilton. The Robin Sage experiment exposes the serious risks created by the uncontrollable impulse of individuals seeking personal relationships. They are lured into the mirage of searching for contacts that may appear to offer an attractive prospect for their personal aspirations, while forgetting all the precautions that a seasoned professional should never discard, whether in the arena of face-to-face encounters, or even more important, in the impersonal encounter with a SNS entity. This entity may or may not represent a real and trustworthy individual. When we are online the dangers of personality dissociation are very real.

The OPSEC protocol is the responsibility of every one entrusted with the handling of sensitive information, but it appears we tend to remain very inclined to drop this commitment when it conflicts with our personal goals. Only as a rhetorical question: Is there anything preventing an adversarial agent from replicating the Robin Sage operation? And correspondingly, are those entrusted with safeguarding the OPSEC protocol committed to comply with the responsibilities associated with OPSEC? Quite frequently, we tend to decide based on the ever present question: "what's in it for me?" So, if we are entrusted with protecting sensitive information while traversing Gigapolis, we must walk with a healthy degree of caution. "We can only ascribe our trust to those who have given us solid reason to thrust them."[36]

Posting a Twitter entry requires an informed decision regarding all the data that is being disseminated, but the great majority of Twitter users are under the erroneous impression that the only thing they are releasing is the text encompassing their thoughts. Any SNS will disclose, to the trained individual, a plethora of data about the individual disseminating that data. Regretfully, most SNS users are either completely oblivious to the meta-data accompanying their posting, or simply they do not care, totally engrossed in using the platform that allows them to become "somebody" in the global cyber village; they have now a podium to show the world who they are and what they think. Is this perhaps the true motivation allowing SNS to enjoy such unprecedented popularity? Articulating this rhetorical question just reminded me of the movie "The Jerk", when the main character sees for the first time his name printed in the phone book. He reacts with great exhilarating enthusiasm while jumping and screaming several times: "I am somebody"!

Do we honestly think that the world cares about our every inconsequential little though and comment that germinates in our minds? Apparently so, because this is the overarching premise behind the posting of our thoughts on every imaginable topic. Once again we live and act according to this erroneous and distorted concept of the democratization of thoughts. We think that every though is equally relevant and worthy the attention of the rest of the world? Really...?

They are lured into the mirage of searching for contacts that may appear to offer an attractive prospect for their personal aspirations, while forgetting all the precautions that a seasoned

---

36 Lou Giannelli, The Cyber Equalizer (Xlibris Corporation, 2012), 35

professional should never discard.

An article in USA Today quotes a courageous female writer who dares to articulate the crux of the matter in the obsessive use of SNS. She states: "We live in a digital age where people are sharing literally every stupid thing happening in their life. But then they suddenly want privacy? If you want privacy, get off Twitter and Facebook ... but if you are online constantly and griping about having your rights infringed upon, I think you are a moron."[37]

This distorted reality has permeated and invaded the working place as well. Enterprise email servers are littered with "reply to all" comments on every single topic, because we want everybody to read what "me, myself and Irene"[38] think about everything, and therefore "me, myself and Irene" will misused the "reply to all" to ensure my opinion is out there in front of everybody. This is clearly a case where a distorted view of CYRE is impacting the work place, because every one considers their unsolicited opinions worth of being broadcast through enterprise-level email systems, clogging the inbox with unnecessary and superfluous traffic, and wasting the enterprise bandwidth. And why? Because we no longer see the enterprise email system as a professional resource, but as an extension of the trivial behavior nurtured by the SNS.

Perhaps one of the most preposterous assertions regarding SNS comes from the founder of one of them. He claims that connecting with people through a SNS site (namely, his) is "a basic human behavior", and he goes so far as to assert that

---

37 Jon Swartz, USA Today, Sunday, December 1, http://www.usatoday.com/story/tech/2013/12/01/tech-firms-counter-nsa-data/3495995/

38 A pun on the 2000 comedy, depicting the behavior of state trooper Charlie suffering from a psychotic disorder resulting in a second personality.

this type of connectivity is a human right.[39] In his poorly written 10-page paper, that of course he posted on his SNS in August 2013, he states that "connecting the world will be one of the most important things" to accomplish. What a perfect example of techno-blindness! So there you have it, folks! All it takes to reach this Internet utopia and make the world better is the get connected to his SNS! The ludicrous idea that the solution to global problems is Internet connectivity was criticized by another influential ICT exponent, who emphatically denied this goal is the most important priority,[40] as advocated by the SNS spokesperson. The criticism to this utopic view is concisely expressed by the Italian article reporting reactions to the 10-page document, addressing the folly of all those "che ritengono internet la panacea di tutti i mali"[41] (considering the Internet as the cure for all maladies).

SNS has become the cradle of a DID-affected generation, where many traditional and secure boundaries of personal privacy and sound ethical principles are being sacrificed in the altar of alter egos.

---

39 Mark Zuckerberg, Is Connectivity A Human Right?, August 2013, https://www.facebook.com/isconnectivityahumanright
40 Iain Thomson, 'It's a joke!' ... Bill Gates slams Mark Zuckerberg's web-for-the-poor dream, The Register, 2 Nov 2013.
41 La Stampa. Tecnologia , Bill Gates contro Zuckerberg, 02/11/2013, http://www.lastampa.it/2013/11/02/tecnologia/bill-gates-contro-zuckerberg-internet-non-salvera-il-mondo-huQpDPrq132PzMnAtcWshL/pagina.html

Chapter 5. Cyber UD

The gullibility and triviality we outlined in the previous chapter discussing SNS provide the foundation for establishing an unwilling but yet effective platform aiding the potential for cyber terrorism. There are myriads of adversarial agents who can simply observe and study our habits and behavior in the cyber dimension, and weaponize legitimate cyber techniques and mutate them into lethal tools for terrorism.

If we do not even understand the cyber reality that surrounds us, how shall we even begin to detect and mitigate the innumerability of factors we create to facilitate the occurrence of cyber terrorism? We may not intend to do so, but we do. Ignorance is not an excuse for creating circumstances where adversarial agents can seed their strategies, and develop their plans to implement cyber terrorist attacks against our critical infrastructures, and place us at risks as a society and as a government.

A recent research paper attempting to elucidate the issue of cyber terrorism concludes there is no unified definition for cyber terrorism in our official lexicon. Thus, the researcher turns to the academia as a source for disambiguation on cyber terrorism, and finds a definition focusing on disruption or destruction of digital property, excluding persons or physical property. This definition highlights the obvious problem: academia may have an intellectual and conceptual understanding of the hyphenated term "cyber-terrorism", but academia does not understand the essence of cyber and the impact of cyber code weaponization. The researcher does exhibit intellectual acuity in stating that academia's definition frequently tends to over simplify the issue.[42]

---

[42] George Jarvis, Comprehensive Survey of Cyber-Terrorism, 28 Nov

However, this researcher negates that very same acuity with his next conclusion. He states that the general consensus is that we have not yet experienced cases of cyber terrorism. His conclusion is faulty since it is based on the argument from silence and the argument from ignorance. Instead, he should have pointed that we do not yet know of any reported cases of cyber terrorism. Lack of awareness or lack of records regarding cyber terrorism does not equate to the absence of such events. As a society we simply may not know about them.

Excursus: Logically fallacious propositions
We human beings are not endowed with the qualities of either ubiquity or omniscience. Our presence is always confined to a single location at any given time, and our cognition is always partial, neither comprehensive nor absolute. For these reasons we should responsibly avoid propositions or conclusions founded on fallacious logic.

Two of the most common logical errors are known as "argumentum ex silentio" and "argumentum ad ignorantiam". The former arises out of the faulty logic of assuming that something not documented corresponds to something inexistent. The latter proceeds from the logical fallacy of assuming that something is false because it has not been proven true.

In the case of the argument from silence we base a conclusion on the absence of historical documentation. The best example of this faulty logic is to affirm that the Great Wall of China doesn't exist because the Marco Polo journal[43]

2011, http://www.cse.wustl.edu/~jain/cse571-11/ftp/terror/index.html
43 Marco Polo, IL MILIONE, Letteratura italiana Einaudi, Edizione di riferimento, a cura di Valeria Bertolucci Pizzorusso, Adelphi, Milano 1975

does not mention it. He was the first European to create a detailed chronicle of his traveling experience to Central Asia and China. However, he doesn't mention this monumental construction.

Then there is the argument from ignorance, where we simply affirm there is no evidence regarding the existence of something. In following this fallacious logic we fail in affirming that because we do not know about something, then that something doesn't exist. This is perhaps the most presumptuous of the two logical fallacies. We are simply saying: "If I don't know about something, then it doesn't exist."

Does the researcher in question know about every case of cyber activity? Does he have access to all the cases of cyber activity? Does he know how to read and interpret binary code? Is he trained in deconstructing and interpreting cyber assembly language? Is he a professionally qualified cyber threat analyst? If the answer to any and all these questions is in the negative, then he cannot, and should not, affirm that we have not yet experienced cases of cyber terrorism. Such conclusion is faulty. It is the result of following the misleading path of argumentum ad ignorantiam.

Neither journalists, nor academia or researchers are in position to pontificate on cyber matters, unless they are unequivocally qualified as cyber matter experts, in possession of the corresponding cyber technical qualifications. There is neither an international nor a national all-comprehensive consensual definition of terrorism. An early research conducted by this author on terrorism[44] discovered there were over a hundred different definitions of

---

44 Schmid and Jongman, Political terrorism: A new guide to actors, authors, concepts, data bases, theories, and literature.

terrorism worldwide, and there are various official American definitions[45] as well. Accordingly, and since the situation is even worse when it comes to define "cyber", there is neither an international nor a national unified definition of cyber terrorism. Yet, at the deepest and primordial level, we all know that terrorism is the type of premeditated and calculated adversarial activity designed to inflict fear and panic on a target population. Consequently, "cyber terrorism is the type of premeditated and calculated adversarial activity designed to inflict fear and panic on a target population by weaponizing cyber code."[46]

A weapon is a device designed to injure or destroy, with the final goal of defeating or killing the opponent. By extension, a weaponized cyber code is a digital device designed to disable or destroy the cyber code of the opponent. However, this is only the first stage; the second stage is the kinetic effect that the disabled or destroyed cyber code will have on the physical systems associated with the affected cyber code. During Black Hat conferences there have been plenty of demonstrations on how to affect systems depending on cyber code attacked by weaponized binaries. Unless cyber SMEs translate these cases to journalists, academia, or researchers, they would never become aware of them because they lack the cyber knowledge to understand the attack, and the physical consequences of such an attack.

When we combine the weaponization of cyber code and the goals of terrorism, as defined in this chapter, then we have cyber terrorism. When the weaponized cyber code has been either compiled or used with a terrorist goal in mind, and the results are the introduction of intimidation, fear, and negative effects on the well being of a targeted population, then we have cyber terrorism in our hands. Imagine the vital water

---

45 Including the DoS, DoD, and FBI
46 Original definition of cyber terrorism by this author. For a definition of the term cyber, consult "The Cyber Equalizer" by this same author.

Cyber Reality

supply for a population, provided by a water purification plant dependent on Industrial Control Systems (ICS). If a terrorist unit finds the way to compile or use a binary code with a malicious configuration, capable of altering the ICS responsible for the proper performance of the purification plant, then we have a cyber-terrorism case in our hands.

Cyber terrorism is the type of premeditated and calculated adversarial activity designed to inflict fear and panic on a target population by weaponizing cyber code.

So let me ask you this. Do you know for a fact that the cyber systems integrated in the ICS are properly protected? And if you answer is in the affirmative, is it because you have empirical data supporting that confidence, or you are simply assuming: "Well, I guess someone is doing that..." Translate the same scenario into the power grid supplying electric power to your area, or the fuel supply sustaining the different industries and transportation complex in your area. All the systems that have a dependency on cyber code, and provide a critical service to a population, are a potential target to cyber terrorism. Are we implementing the responsible and effective cyber protection required by these critical systems? Do the people administering these system have a contingent of certified and empirically qualified cyber SMEs to maintain a proactive monitoring cyber defensive program? Up to 80% of the victims of cyber attacks became victims because of apathy or incompetence.

It is important at this junction to differentiate between the use of cyber systems to perform an action, which may contribute to a greater terrorist plan, and the use of cyber systems to subvert other cyber system, for the purpose of inflicting harm and terror on the target population.

Cyber Reality

For example, cyber systems may be used to purchase a plane ticket on-line. This ticket may be used by a terrorist to travel to the terrorist target, but the action of buying the plane ticket is not an act of terrorism in itself. A case of cyber terrorism occurs when cyber systems themselves are the target, and the intent is to modify or disable them in order to introduce destruction, chaos, fear and panic in the target society depending on those cyber systems.

With so many multiple cyber identities collected and used by individuals employed in sensitive and critical areas, the theft and misuse of such cyber identities offer what is perhaps one of the most insidious aids for cyber terrorism. Let's consider the following feasible scenario, facilitated by the convenience of online stock trading. An influential and trusted cyber stock trading identity is stolen and stocks are manipulated in order to introduce financial destabilization in the trading market. The motivation for such a potential cyber terrorist attack may have a simple or a combined purpose: to destabilize the market for the direct result of economic chaos, or to profit from the stock manipulation in order to finance ulterior terrorist acts.

So, when interacting in the SNS arena, we must exercise extreme caution, and realize that our careless activities and associations may unwillingly contribute to create factors and circumstances that, once identified and exploited by cyber adversaries, may allow them to further their terrorist plans and operations, against our nation, or any other nation.

Let's remember that the cyber spectrum is nonlinear,[47] and what may appear as a trivial or insignificant action, may result in an exponentially larger consequence. Since it is unfeasible and impractical to expect the cyber population at large to implement personal cyber security measures among

---

47 Giannelli, 109

cyber users, the most we can expect is a sense of general caution and alertness. The path to grasp the true essence of CYRE begins with the small but significant step of changing our perception of the cyber arena: it is not an entertainment sandbox, but a dangerous environment, so much so because of the degree of pseudo anonymity associated with walking and traversing this cyber terrain.

Why is this commitment to caution and alertness so important in the cyber community at large? Because the one factor that enables cyber terrorism above all others is a target of opportunity. This type of target may emerge at any time, and once it becomes evident, a cyber terrorist will focus on it and will take the appropriate steps to exploit it. A target of opportunity is quickly detected in the cyber dimension, since it eradicates the distance barrier. It is certainly not a hyperbole to say that the discovery of a target of opportunity is like a drop of blood in an ocean area populated by sharks.[48]

Perhaps one of the greatest challenges we will face as we start structuring a framework to assist us in dealing with cyber terrorism will be very similar to the problem of networking and cyber security. In our zeal to sink our teeth into the problem we oversimplify the task in front of us, and we fragment the issue in two, instead of dealing with the synthesis of the issue.

Those who have been dealing for a long time with networking issues tend to believe that cyber security is just a footnote in networking history. Likewise, those dealing with terrorism will tend to assume cyber terrorism is just another footnote in terrorism history. The problem with this oversimplified approach is best illustrated by the myriad of

---

48 According to a National Geographic article, some sharks can detect minute amounts of blood from more than 2 miles away. See http://animals.nationalgeographic.com/animals/fish/great-white-shark/

cyber security issues affecting all those who consider that cyber security is only an afterthought in networking communication, and their indolent behavior actually facilitate the arrival of the never ending series of cyber exploits.

Likewise, those who may approach cyber terrorism as just an afterthought of terrorism will equally invite a plethora of opportunities offered to the adversaries who will use these cyber means as a powerful enabler for their cyber terrorist exploits. So, the sober reminder that should remain as a constant alarm pulse is: what am I doing that is contributing as an enabler for cyber terrorists to exploit my personal or my enterprise data?

Here resides the crux of the issue: the creation of opportunities to bestow our adversaries with numerous opportunities to exploit our cyber data for adversarial purposes. Three main components comprise the core of the problem: the weaponizing of cyber code, the unauthorized extraction of data, and the consequences of the unauthorized data extraction. In the following paragraphs we will focus on details pertinent to these three components.

Any cyber code is potentially a candidate for weaponization. Code weaponization (COWEP) is the adversarial technique, with either an internal or external agent, of modifying cyber code designed originally with a benign purpose, and mutated into a malign cyber code. The intent of COWEP is to disrupt, modify, or destroy a single cyber system, or a cluster of them. When COWEP is used to negatively affect cyber systems that are responsible for maintaining a critical function in support of a given society, then the final purpose mutates into cyber terrorism, to harm that targeted society by introducing panic, fear and destruction.

Cyber Reality

The unauthorized extraction of data[49] is reaching pandemic proportions due to the casual attention to the operations security (OPSEC) protocols and network security (NETSEC) guidelines. The transfer of digital data, whether authorized or not, is a trivial process in the cyber dimension. However, the effects and consequences of unauthorized data extraction (UDE) can have enormous ramifications, covering personal, financial, political, and national and international repercussions.

What am I doing that is contributing as an enabler for cyber terrorists to exploit my personal or my enterprise data?

The only way to ensure that significant, important or critical data maintains its ICA status is to design, implement and enforce the appropriate OPSEC and NETSEC protective measures. The ICA status refers to the foundational principles on information security (INFOSEC), encompassing the triad of conditions required to ascertain the data is secure, when and only when we can ensure the integrity, confidentiality, and availability (ICA) of the data under our care.

By maintaining the ICA status of data we ascertain that the data has not been modified by any unauthorized agent. Thus the data retains its original and protected integrity. The

---

49 This author refuses to use the spurious term "exfiltration" popularized by certain circles who think that an antonym is automatically formed by exchanging the prefix "in" for "ex". Infiltration is a proper English term dating back to the 18th century, as opposed to the spurious counter-term carelessly coined in late 20th century writings. I can assure you that performing the same misleading prefix exchange will not result in creating a proper antonym. Would you like to try with the term "increment" for instance?

confidentiality requirement is the assurance that only those with the proper authorization and need-to-know are granted access to the protected data, thus retaining its confidentiality. Finally, the requirement of availability ensures that the data remains accessible and available when the authorized agents are granted access according to the mission requirements. The following figure illustrates the ICA triad.

But what do we really do when it comes to data transfer? We pay lip service to OPSEC, NETSEC, and INFOSEC, and in the process we manage to add indolence, ignorance, and even impudence into the mix. How may you ask? Let me tell you...

In addition to all the maladies we spread among the users in the cyber dimension, we have an additional one, generated and spread by the impertinent agenda of libertarians who self-appoint themselves as the spokesperson of the population. They abuse the launching platforms of their media outlets, and they publish information leading to enabling terrorism in general, and cyber terrorism in particular. And on top of all this, they bestowed upon themselves Pulitzer Prizes on a regular basis, as a reward for their perfidious actions. Let's examine a few cases, in chronological order, from our recent history.

The American Cypher Bureau began its operation in New York, and its best known work took place during the Washington Conference of 1921-22, when this unit decrypted the cipher used by the Japanese representatives attending this conference. This decrypted data enabled US negotiators to obtain better terms from Japanese delegates. Despite this success, among several others, the unit was dissolved in October of 1929. Yardley, the Chief of the Cyber Bureau, now unemployed, decided to write about his cryptanalysis operation, first in a series published in The Saturday Evening Post, and later, in 1931, in the book The American Black Chamber.[50]

This is the first documented case of unauthorized disclosure (UD), though not the last one. There have been many other cases, equally or even more damaging to the US interests and US safety. And who may I ask, rhetorically speaking of course, benefited from this UD? The Japanese of course. After learning from the press coverage and the book, and the revelations compromising the sources and methods used by the American cryptanalysis unit, the Japanese changed their codes and encryption techniques, and placed US cryptanalysts in a disadvantageous position. Because Yardley had not technically violated any existing law at the time of his UD, the USG did not prosecute him.

With this historical fact in mind I have another rhetorical question for the reader: Do you suppose the history of World War II would have been different if the US had maintained the cryptanalysis advantage we had gained

---

50 National Security Agency, Central Security Service, Pearl Harbor Review - The Black Chamber, Herbert O. Yardley, http://www.nsa.gov/about/cryptologic_heritage/center_crypt_history/pearl_harbor_review/black_chamber.shtml; U. S. Naval Institute, http://www.usni.org/store/books/history/american-black-chamber

Cyber Reality

before the UD of the Black Chamber?

Then, another case arises form the Cold War era. In 1968 a Soviet submarine, carrying nuclear-armed ballistic missiles, sunk to the Pacific ocean floor. The US located the submarine about 1,500 miles northwest of Hawaii, at a depth of 16,000 feet. Under the support of the US DoD, the CIA designed and launched Project Azorian, involving a secret six-year plan to retrieve the sunken Soviet submarine. Project Azorian represented an unprecedented treasure trove of Soviet intelligence. A special recovery ship was built, under the ownership of billionaire Howard Hughes. The cover story stated the ship Glomar Explorer was designed for marine research on extreme depth mining operations on the sea floor.[51]

The recovery operation began in 1974 under secrecy, despite the monitoring of nearby Soviet ships. A giant underwater claw grabbed the hull of the Soviet submarine, and initiated the long ascend from the ocean floor. Regretfully, the submarine hull broke apart when it was a third of the way up, and a section plunged back to the ocean floor. However, the Glomar crew was successful in recovering the portion that remained in the recovery device, and the planning for a second mission to recover the lost section began immediately. But once again the press intervened and frustrated the plan by implementing yet another case of UD.

In June 1974, just before the Glomar left port, thieves stole secret documents from the offices of the Summa Corporation.[52] The stolen documentation established the nexus between Howard Hughes, the CIA, and the Glomar

---

51 Central Intelligence Agency, Project AZORIAN, https://www.cia.gov/about-cia/cia-museum/experience-the-collection/text-version/stories/project-azorian.html
52 The contemporary name of the Howard Hughes corporation

Explorer. The FBI enrolled the Los Angeles Police Department to recover the stolen documents, but the search efforts captured the attention of the media.

Despite the personal appeal from the CIA Director requesting the media not to disclose information about the Project Azorian, on February 1975 the Los Angeles Times published information connecting the robbery, Hughes, the CIA, and the recovery operation. Following this initial disclosure, reporter Jack Anderson revealed the story on national television. OK, we already know the drill by now, so let's ask ourselves the question: who benefited from this UD? Certainly not the American intelligence, but the Soviets! They assigned naval forces to monitor and guard the recovery site, thus forcing the White House to cancel further recovery operations.[53]

The cost of Project Azorian is estimated at $550 million, but after the UD the Glomar was disqualified as a platform for future intelligence operations. What were the losses in terms of maintaining our capability to recover Soviet intelligence? Even greater than the monetary losses, courtesy of the impertinence of the American media!

On February 2010 the CIA published an article on the Glomar Explorer project. A comment to this release states that the CIA document greatly elucidates what the public knew "about this poorly-understood operation."[54] This is certainly a very misguided and irrelevant comment. Why is it so vital that the public knows and understand about this intelligence mission? The public alluded in this comment is not part of the secret Project Azorian. Therefore, they do not

---

53 Ibid
54 Matthew Aid, William Burr and Thomas Blanton, editors, The George Washington University, Project Azorian . The CIA's Declassified History of the Glomar Explorer , The National Security Archive , February 12, 2010 ,
http://www2.gwu.edu/~nsarchiv/nukevault/ebb305/

have a need to know.

Furthermore, the same comment laments that the CIA article does not answer "the critically important questions" regarding the size of the recovered portion of the submarine, or the amount of intelligence extracted from the recovery. What impertinence and petulance! Critically important for whom? Once again, this is a mission conducted by the US intelligence, for the purpose of collecting intelligence on the adversary, and it is a secret mission. The media and the general public are not part of this operation, and therefore they do not have a need to know about the results of this operation.

US counterintelligence has the difficult mission of preventing both external and internal spies from obtaining information on our national intelligence. Most recently foreign spies have had no need to design and launch complicated and dangerous procedures to steal our national secrets; all they have to do is to wait for Americans to place them in the hands of unscrupulous journalists. And the arrogance of these journalists compel them to publish data from UD cases despite of government pleas for discretion.

A reputed publisher of The Washington Post during a public speech in 1986 acknowledged that on occasions the media has made tragic mistakes. She illustrated her point by citing the UD case of revealing that American intelligence was collecting coded radio traffic between terrorist in Syria plotting with their overseers in Iran. After the American media published this information and alerted the terrorists, the radio traffic stopped, and five months later they attacked and destroyed the Marine barracks in Beirut, killing 241 Americans.[55]

---

55 Scott Shane, A History of Publishing, and Not Publishing, Secrets, The New York Times, July 2, 2006, http://www.nytimes.com/2006/07/02/weekinreview/02shane.html?

We also have the the documented case of the unstable individual that enlisted with the US Army and obtained a high level clearance, only to misuse it in order to gain access to restricted information and place it in the hands of unscrupulous journalists. Chelsea Elizabeth Manning, formerly known as Bradley Edward Manning, was convicted in July 2013 of violations of the Espionage Act, after releasing the largest UD set of classified documents to the public. The cache of UD documents included 250,000 US diplomatic cables and 500,000 Army reports, with the bulk of this restricted material published by WikiLeaks and its media partners during April and November 2010.[56]

Reporters Without Borders criticized the court decision, portraying the 35-year prison sentence as "disproportionate", and proclaiming the Manning case illustrates the vulnerability of whistleblowers.[57] A member of the US Armed Forces has legal responsibilities, and is accountable before a court of law. A member of the US Armed forces willingly misusing the privileges of having a security clearance by purposely committing an act of UD is not a whistleblower; is a traitor.

There is an incisive recent book, and an equally penetrating article exposing the "self-interested complacency" of journalism. The author of this article remarks: "What kind of a First Amendment do citizens and journalists need if they are to undertake the work of democracy? They need something more than a license to make money and to turn the political system into a commercial arena for profligate advertising and consumption of politics."[58]

_r=2&oref=slogin&
56 http://en.wikipedia.org/wiki/Chelsea_Manning
57 Reporter Without Borders , Lengthy prison term for Bradley Manning , 21 August 2013, https://en.rsf.org/united-states-lengthy-prison-term-for-bradley-21-08-2013,45087.html
58 James W. Carey, Journalism and Democracy Are Names for the Same Thing, Nieman Reports,

As per the media announcement on 14 April 2014, the Pulitzer Prized was awarded to The Guardian and the Washington Post "for public service journalism"[59] involving the publication of the UD orchestrated by the former NSA contractor, now hiding in Russia. The Pulitzer committee stated they decided to confer the award in recognition of the manner in which the media reports assisted the public in understanding how the disclosed data correlates into the national security framework. And where in the Constitution of the United States says that the public in general and the media in particular need to interfere into the national security framework?

The people elect representatives to appoint leaders charged with the responsibility of administering and maintaining the core areas of government. If these appointed leaders discharge their duties in accordance to the corresponding laws, then the public is not to interfere with the particulars pertaining to these areas of government. National security is one of the core areas of government. Thus, neither the press nor the public have any business in demanding access to data pertaining to the national security framework.

National security activities must be conducted in secrecy, in order to control the outcome of the plans outlined by the leaders appointed to that mission. The idle curiosity of our population is well illustrated by Mr. Brokaw's question to Gen. Alexander about NSA's secrecy. Mr. Brokaw asked why the agency could not reveal even the number of its employees. And my response to this impertinent question is: why does the people of the US need to know that information? That information is irrelevant to the well-being

---

http://www.nieman.harvard.edu/reports/article/101943/Journalism-and-Democracy-Are-Names-for-the-Same-Thing.aspx

59 BBC News, Washington Post and Guardian share Pulitzer Prize, http://www.bbc.com/news/world-us-canada-27029670

of the US population, but it's extremely valuable to our adversaries. Is there a way to tell the American people without telling the terrorists? Obviously not.

Do journalists understand the principle of information aggregation? Of course they do! Most of their journalistic work operates on this principle. The information aggregation principle states that data from many sources can lead to the acquisition of more information transcending the sum of the collected data units. Deductive reasoning can allow our adversaries to acquire a very clear picture of a target by correlating data units apparently disconnected. Answering the question posed by Mr. Brokaw will result in facilitating intelligence to our adversaries.

The concept of "transparency" exists in connection to information that is relevant to the well being of the population of a country, and it should be publicly disseminated, in a transparent manner. On the other hand, information regarding national security is not public information since it must be protected from exposure, and made available only among those elected and selected with the "need to know" principle.

The town criers should learn their place in society and in democracy. No journalist has ever been elected or appointed as the watch dog of democracy. The media should learn its place as news tellers, the kind of news that are part of the information flow concerning the well being of the population in general. National security affairs is none of their business. Handing out Pulitzer Prize awards does not legitimize unlawful disclosures. It only indicates the presence of a distorted sense of ethics and allegiance to our nation.

The cost of the UD orchestrated by the fugitive NSA contractor is staggering. He is charged with espionage in the US, and I refer to him as the "S-criminal" because I am not going to stain the pages of my book with his name. With premeditation he abused his clearance to gain access and download almost 2 million restricted documents, including data on the identity of undercover operatives and Pentagon programs, forcing extensive changes to all branches of the US Armed forces, caused by the all-inclusive compromised data.[60] This places our national intelligence programs and assets at a great disadvantage, and danger. The goal of the NSA collection programs was to detect and invalidate the planning and execution of terrorist attacks against the US. Now that terrorists have the information the S-criminal unlawfully disclosed with the assistance of the media, they will adjust and modify their plans, and the corresponding gaps in our intelligence on their modified terrorist plans will cause some of our people to die. Furthermore, the negative effects of this UD case will also impact our capabilities to protect civilian targets against cyber attacks.[61] The media orchestrating this UD case, on the other hand, thinks all this is worth a few Pulitzer Prize awards!!

The harm inflicted to our nation is quantifiable. The UD case at stake has contributed to the enabling of the persistent and blatant ongoing espionage and cyber operations from Russia and China, and a variety of threats from al-Qaida and other terrorist groups. They have truly benefited from the disclosures of our national intelligence sources and methods, facilitated by the S-criminal who remains a fugitive in Russia, avoiding temporarily the charges of espionage

---

60 Kimberly Dozier and Stephen Braun, Snowden leaks lead to Pentagon costly change, The Associated Press, http://www.navytimes.com/article/20140204/NEWS05/302040015/Official-Snowden-leaks-lead-Pentagon-costly-change
61 David E. Sanger, N.S.A. Director Says Snowden Leaks Hamper Efforts Against Cyberattacks, The New York Times, March 4, 2014, http://www.nytimes.com/2014/03/05/us/politics/spy-chief-says-leaks-hamper-protection-against-cyberattacks.html

and stealing government property brought against him.[62]

The premeditation of the S-criminal is evident in his interviews announcing that he applied for a position within NSA with the purpose of gaining access to restricted material he wanted to extract. The extracted data contained high classification plans to counter Chinese cyber attack capabilities, and detailed budget information on NSA operations. This data will allow China to obfuscate future cyber attacks and other threats, and will substantially diminish our ability to protect our people, industry and operations.

Espionage is a constant threat, but the capabilities afforded by digital data storage has multiplied this threat. The amount of data extracted by the last two cases covered in this chapter are truly unprecedented. And yet, our general behavior toward digital data and CYRE remains customarily trivial, thus facilitating careless handling of sensitive data, and enabling those bended on betraying our country with internal espionage.

---

[62] Nick Simeone, Clapper: Snowden Caused 'Massive, Historic' Security Damage, U.S. Department of Defense, American Forces Press Service, Jan. 29, 2014, http://www.defense.gov/news/newsarticle.aspx?id=121564

Chapter 6. The Legal Aspect of UD

When it comes to the legal responsibilities of an individual employed by the USG, whether a member of the Armed Forces (military or civilian), or a contractor, there are specific written laws regulating how these individuals are to handle restricted information pertaining to their duties with the USG. In this matter there is no room for personal opinions, whether from the public in general, or the media in particular. When an individual accept employment with the USG, that same individuals enters into a legal contract with the USG, attested by a signed legal document designated as a non-disclosure agreement (NDA).

This NDA is officially designated as STANDARD FORM 312 (Rev. 7-2013), as prescribed by the Office of the Director of National Intelligence (ODNI). The title of the SF 312 reads: "CLASSIFIED INFORMATION NONDISCLOSURE AGREEMENT. AN AGREEMENT BETWEEN [the US person] AND THE UNITED STATES."[63]

This SF 312 is extremely important to properly understand the legal obligations of the person accepting this special NDA. In paragraph 1 the USG employee states this document's intent is to be legally bound, and the signing party accepts the obligations outlined by Executive Order 13526, and other Executive statute prohibiting the unauthorized disclosure of information in the interest of national security.[64]

Paragraph 2 states the acknowledgment of receiving and understanding the security indoctrination outlining the nature

---

63 The SF 312 is available from the GSA site
    http://www.gsa.gov/portal/forms/download/116218
64 STANDARD FORM 312 (Rev. 7-2013), July 2013

Cyber Reality

and protection of classified information, and the procedures to be followed in ascertaining whether other persons are officially approved to receive this restricted information. In paragraph 3 the USG employee acknowledges that the UD, unauthorized retention, or negligent handling of classified information could cause irreparable harm to the US. The signing party therefore agrees to never divulge classified information to anyone not properly authorized by the USG and with an official need to know status.

Paragraph 4 states the signing party has been advised that any UD of classified information may constitute a violation of the US criminal laws, subject to prosecution. In paragraph 7 the signing party acknowledges that failure to return the material comprised in the UD is also a violation of the US criminal laws.

In paragraph 8 the signing party states that all the conditions and obligations imposed by the NDA apply for a life time, beginning at the moment of being granted access to classified information. Paragraphs 10 and 11 state that the provisions of the NDA accepted by the signing party do not conflict with any whistleblower protection, the Military Whistleblower Protection Act (governing disclosure to Congress by members of the military), or the Whistleblower Protection Act of 1989 (governing disclosures of illegality, waste, fraud, abuse or public health or safety threats).[65]

The conditions and provisions of the SF 312, a requirement for USG employee being granted a clearance to work with classified information, clearly expose all the misinformation fed to the public by the media. Neither Manning nor the S-criminal are private citizens conducting volunteering work. They are USG employees bound by the legal contractual conditions they accepted when they signed the SF 312, and they know they hold a position of trust prohibiting them from

---
65 Ibid

engaging in UD. Both of these criminals knew that the media is not authorized to receive classified information, and they knew that divulging this restricted information constituted a violation of the US criminal laws. They acted with premeditation, by willingly exposing the USG to grave harm!

Manning is already serving imprisonment, but the S-criminal is still a fugitive, willingly refusing to return the classified material he unlawfully stole. The restricted information they exposed does not fit into the description of the provisions to protect whistleblowers. Accordingly, they are not whistleblowers, but rather willful criminals and traitors.

The public must also know that even the USG information marked "unclassified" is not automatically approved for public release. Such information may be especially sensitive, subject to the handling procedures established by US laws. According to Executive Order 13526, unauthorized disclosure is defined as "a communication or physical transfer of classified information to an unauthorized recipient."[66] Therefore, any instance of USG information unlawfully placed in the hands of the media (an unauthorized recipient) constitutes a UD case, and a violation of US laws.

This is the legal environment in which any USG employee operates. But what about the legal milieu in which the media operates? First, in a democracy such as ours we need to establish a delicate balance between the right of the public to information relevant to the well being of society on the one hand, and the right and need of the government to maintain and safeguard sensitive and classified information from unauthorized disclosure on the other hand. So the media

---

66 The White House , Office of the Press Secretary , Executive Order 13526- Classified National Security Information, December 29, 2009, http://www.whitehouse.gov/the-press-office/executive-order-classified-national-security-information.

mantra about absolute transparency is just one of their numerous schema to gain access to information that may advance their own self-serving goals. The more disclosure of information pertaining to restricted, confidential, and classified matters, the better for their business. If they hurt national interests, or endanger lives in the process of disclosure, that is not part of their equation. Their self-serving goals is all that matter to them.

A courageous lady, and former head of US Counterintelligence, spoke publicly on September 2013 regarding the harm caused by the two most damaging UD cases discussed in this chapter. She also addressed the role of The Guardian and The Washington Post in exposing the data stolen by the S-criminal. She stated there has been no discussion of prosecuting these two media outlets. However, she reminded the audience that Title 18 of the US Code Section 798 is the statute pertaining to the espionage laws that criminalizes the publication of classified information, and consequently, they could be prosecuted.[67] The very sad reality in which we currently live is that when confronted with the choice between national security and the distorted understanding of freedom of the press, we always sacrifice the former on the altar of the media goddess.

She stated that lawyers can get disbarred for violating their ethical obligations. So where are the media ethical standards holding them accountable for violating laws protecting our national security? Of course, the media is going to reply that The Society of Professional Journalists has adopted a Code of Ethics, but it is not binding for journalists, and it's offered only as a guide for making ethical decisions. This code is completely silent on matters of conscience or legal obligation regarding the protection of national security secrets, and

---

67 Michelle Van Cleave, Myth, Paradox & the Obligations of Leadership, September 13, 2013, http://www.centerforsecuritypolicy.org/wp-content/uploads/2013/10/Van-Cleave-Occasional-Paper-1011.pdf

void of any considerations for lives placed at risk by the unlawful disclosures disseminated by the media.

Let's briefly consider a couple of examples about the issue at stake. Stanislav Lunev, former Russian military intelligence officer and author, expressed his amazement, along with Moscow's appreciation, at the numerous times when he found very sensitive information in American newspapers. In his opinion, he sees Americans as caring more about competing with each other than protecting national security.[68] Lyle Denniston, a journalist for the Baltimore Sun, summarized his ethical standard with these words:

"As a journalist, I have only one responsibility and that is to get a story and print it. It isn't a question of justification in terms of the law, it's a question of justifying it in terms of the commercial sale of information to interested customers. That's my only business. The only thing I do in life is to sell information, hopefully for a profit."[69]

The former head of US Counterintelligence explained in very plain language the US intelligence plan of collecting phone metadata for the purpose of tracking the plots of terrorists and adversaries in general. Her words are worthy of a verbatim citation.

"The central decision behind the U.S. telephony metadata collection program (Sec. 215 of the Patriot Act) is straightforward commonsense. If you want to know who the terrorists are talking to, you've got to check the phone logs. It's dot-connecting 101. We need to search for terrorists'

---

[68] Unauthorized Disclosures of Classified Information, The Office of the Director of National Intelligence (ODNI), Office of the National Counterintelligence Executive (ONCIX), September 2011, http://www.ncix.gov/training/WBT/docs/UDB_091211.pdf
[69] Ibid

footprints in order to prevent further terrorist attacks. Is your phone not involve in this footprint? Then what is your problem? If you don't have the responsibility of guarding national security, then let those who do hold such responsibility do their job, and go on with your life. Who is inflaming these sentiments among the public? The press, with their irresponsible choice of publications. All they care is satisfy idle curiosities, and make money in the process of doing so."[70]

---

70 Ibid

Chapter 7. Global Cyber Identity

In the world of programming is very typical to start the first lesson with an exercise that allows the student to measure one's assimilation of basic principles of programming by writing the code that, when executed, will render the message "Hello World".

Those who become users of the so-called "social network" soon succumb to the very alluring compulsion of posting as much information about themselves as possible, their personal and professional lives, to the point of reaching a level that quickly become a digital overexposure, by revealing aspects of their lives that they wouldn't dream of committing to paper. The social network becomes their way of saying "hello World" to the entire global digital village.

This pathological compulsion for digital exposure reaches dangerous levels, and the posted information becomes a threat that eventually will target the very same person who posted the information, and the safety of those associated with the individual posting information on the social networks. The threat of identity overexposure is categorized in two main categories: intended and unintended

A cursory review on the topic of social networking safety will reveal that a majority percent of concern focuses on the topic of online child safety. While this fact highlights the importance of protecting children from online dangers, it also highlights the misguided fallacy of assuming that adults may not be as vulnerable as children when facing online activity in general, and social networking in particular.

The very nature of the global presence in the cyber dimension renders what we perceive to be our local cyber identity into a global cyber identity. Why? Because we tend to hold the misconception that cyber activity is a local activity. The big lesson for us to learn is that the nature of cyber data is intrinsically global, because the cyber dimension is global. The data we place into the cyber highway is not restricted by geographical or international boundaries. It literally travels around the world.

We may, of course, choose between establishing a genuine cyber identity, or an anonymous one. This latter option is taken when we desire to either protect our privacy by dissociating our persona from our "cyber persona", or to attempt to evade accountability for our cyber activities. However, the anonymity option is not a real option; it's only a contingency. Our true identity will become exposed when a systematic, methodical, persistent, and technically proficient agent knows and employs the numerous unique identifiers associated to our cyber persona. Those identifiers are available to the agent with the cyber technical knowledge required to unveiled the connection between our persona and our cyber persona.

Whenever we use a cyber device, at work, at home, or on the go, if we establish network connectivity, consequently and inevitably we will leave a bread crumb trail, pointing to the source of the network data we originate. Granted, there are a number of ways through which we may modify these identifiers in order to obfuscate the origin of the network data stream. Some of them are easier to modify, and other more difficult, but we cannot modify or obfuscate all of them. The vast majority of the pseudo cyber detectives attempting to establish attribution for a certain network data stream speak of tracking the IP address that points to the point of origin. Then they proceed to pontificate regarding the identification of the alleged source.

This oversimplified and amateurish approach reveals an

overwhelming amount of lack of understanding on how network data streams traverse a number of networks. In our global setting of network connectivity the journey of the network data stream is usually very complicated. In determining attribution we must be very cautious not to employ the services of cyber neophytes, a very prolific breed nowadays. Network source attribution is perhaps one of the best illustrations to the axiom that a little knowledge is dangerous. If we do not have a mature and seasoned technical and empirical knowledge of networking we are in danger of misinterpreting the network data stream, and arriving at the erroneous attribution.

The cyber data generated at the cyber system standing as the true source will traverse not one, but a rather large number of networks between source and destination. The cyber data at the point of origin is prepared and arranged for the long journey by being divided in small segments,[71] arranged according to a predetermined order, and marked with a series of identifiers designed to facilitate their journey and prepare them for reassembly at the final destination. This stream of data, now divided in a series of data segments, is directed to traverse a significant number of intervening networks between the source and destination. Every intervening network will also place additional identifiers to the traversing data segments, in order to facilitate the orderly progression of these segments.

Someone may say: "Well, I know that every data segment contains information about the source IP address, and the destination IP address." That affirmation is correct. And then, encouraged by this initial success, that someone may add: "So, when I read the source IP address in the data stream I know the IP address of the system originating the

---

71 In networking parlance we refer to this data segments as "packets", assuming we are using the most popular TCP/IP suite of networking protocols.

data stream, and I can determine attribution." This affirmation is incorrect in the vast majority of cases. It would be correct only in a minority of cases where the data stream traverses only a single internal network, and that's rarely the case.

When a data stream traverses the Internet from the source to the destination, all the corresponding data segments (packets) will travel from one router on an intervening network, to another router on the next intervening network. This process will be repeated as many times as necessary in order to travel the entire distance between source and destination. Very, very rarely the origin and destination reside in the same network.

Therefore, the process of determining attribution is complicated by the complex journey of each data packet traversing a large number of intervening routers. When the data leaves the cyber device of origin, it goes to the router on that local network. This router then calculates, based on internal routing tables, what is the next intervening router in the long series between the origin and destination networks. The packets are not directly sent to the destination, but rather to a series of intervening networks, and their corresponding routers, responsible for routing the data packets to the next router on the long journey to the final destination.

Every intervening router advances the packets to the next hop (router) in the selected route toward the final destination. Every intervening network may be a complex one, with several intervening routers and associated homogeneous networks. A complex of homogeneous network sharing a common routing policy is known as an Autonomous System (AS), and as such, it is registered with a unique identifier, known as the AS number (ASN). Thus, every AS and its unique identifier differentiates that particular AS from the neighboring distinct homogeneous networks, with their own corresponding ASNs.

Where are we going with all these explanations about the route followed by a data stream and its component packets? We are establishing the framework that theoretically allows the tracking of any particular data stream, because every intervening network has a unique identifier, thus providing the foundation for establishing a tracking methodology allowing to establish attribution, regardless of whatever attempts to disguise the identity and location of the true source of the data stream. The real difficulty in establishing attribution resides in the complex task of gaining access to all required records to establish attribution, but such records exist. So, it is difficult to establish attribution, but certainly not impossible. The difficulties are not technical, but rather logistical and legal in nature.

Every SA belongs to a particular organization or entity, and the agent attempting to establish attribution may or may not have, or being granted, access to the network records of the many intervening AS holding the records showing the itinerary of the particular data stream under scrutiny. So for those who intentionally attempt to hide the true source of a given data stream, their bluff will work only until the scrutinizing agent receives access to the records of the intervening AS. This has been done, and continues being done, whenever the case under scrutiny receives the attention and the cooperation of all the AS involved in the transit path of the data stream in question. The traversing path, followed by the data stream, is recorded, and a qualified investigating agent can obtain this data, provided this agent is equipped with the knowledge and the necessary contacts required to obtain this information.

Let us use a simplified diagram of a routing and forwarding networking path from source to destination, with intervening

Cyber Reality

ASNs, to illustrate the journey of a particular networking data stream.

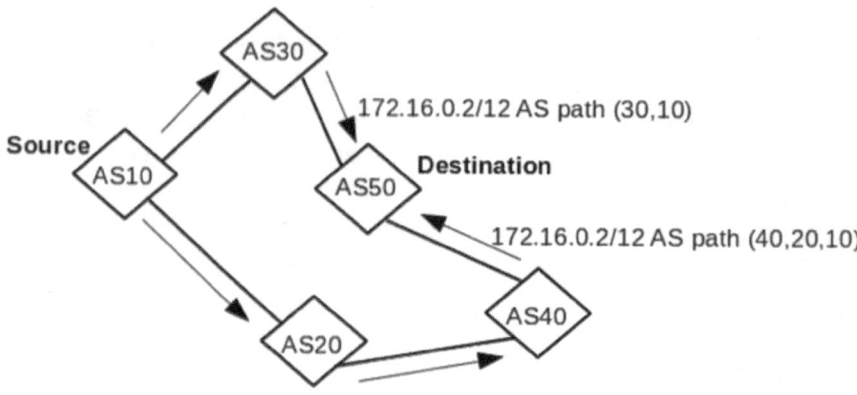

The record of the path followed by the data stream packets is indeed available. The source cyber device generating the data stream resides inside a particular AS, and when the data stream traverses the neighboring AS (according to the logical distribution of the available AS), the path formed by the ASN forwarding the routed data to the destination becomes part of the record. Consequently, when the packets arrive at the destination ASN, the routing advertisement data contains the information on the entire routing path. In the example provided in the graphic above we have the source ASN10, and the destination ASN50.

The packets from the original data stream may have been routed through different routing paths, but when they arrive at ASN50, the record shows these packets originated on the network 172.16.0.2/12. One of the divided streams followed the path originating at ASN10 through ASN30. The other divided stream followed the path originating at ASN10, through ASN20 and ASN40. The routing advertisement contains information resulting from the "stamping" taking place at every intervening ASN, where the current ASN

Cyber Reality

prepends its ASN to the preceding AS path of the arriving data stream. At any given time the traversing data stream shows the sequence of connected ASNs traversed between the current point and the source ASN.[72] This information offers a tremendously important tool for source detection and attribution.

Let's return momentarily to the neophytes who think they can oversimplify the task of establishing attribution by tracking the information on the source IP address. A person determined to obfuscate the true source of a data stream can simply evade detection by compromising vulnerable cyber devices and use them as proxies, as their platform for launching their malicious cyber activities. Do the owners of those compromise systems being used as proxies know their systems are compromised? Do they have the technical skills to read the cyber raw data showing they are being used as proxies by a malicious cyber entity? This type of scenario is real and very common, but even though these occurrences are recognized as an international issue of concern, there is no internationally coordinated legislation allowing authorities, local or international, to pursue the task of establishing attribution, and legal corrective action, when merited.

Another factor in the difficulty for establishing attribution is the manipulation of the data stream packets. Each data packet routed through the Internet contains information regarding its source and destination, along with other unique identifiers. However, the information in the source field in the data packets can be changed, or spoofed, by the cyber entity seeking to conceal the true source. The spoofing action simply makes the stream data appears as if coming

---

[72] Geoff Huston, APNIC, CISCO, Exploring Autonomous System Numbers, http://www.cisco.com/web/about/ac123/ac147/archived_issues/ipj_9-1/autonomous_system_numbers.html

from another system, selected either randomly, or by design. Further complications in establishing attribution arise when the nefarious cyber activity employs a botnet, a collection of compromised cyber devices acting under the command of an attacking entity. This technique not only amplifies the strength of a cyber attack, but conceals even further the identity of the true source. Thus, an oversimplification on the process of establishing attribution may very easily lead to point an accusative finger to a spoofed source, not the true source.

You may ask: But why do we allow that capability, facilitating the concealing of the true source, and consequently, the identity of the malicious entity behind this malicious manipulation? Because when we designed the TCP/IP suite of protocols now so prevalent in the Internet network traffic, we were still leaving in the age of innocence,[73] when we thought everybody joining the Internet was a trustworthy persona, seeking only to share information over the Internet, with good intentions. We completely overlooked the dark reality of human malice and selfishness, now the driven force for every nefarious cyber action plaguing the global Internet village.

Tracking down a particular data stream to its source it can be done, but there is also a dimension of interpretation of the data we are tracking back to its true source. An over simplistic level of knowledge of networking, and an amateurish degree of technical skills are not sufficient to confidently track the source of a data stream. However, not even an advanced level of cyber knowledge, skills and technical means can truly pinpoint the true source of a particular data stream. The only way to achieve a degree of high confidence in attributing the source of a data stream is by fusing technical means and HUMINT.[74] The detection of

---

73 Giannelli, chapter 3
74 Human intelligence, the system of collecting information via human

Cyber Reality

the true source for a data stream that represents a threat to our personal or enterprise identity and operational security is best achieved only when we fuse in a synergistic manner the information proceeding from CYBINT, HUMINT, and SIGINT.[75]

Our personal and cyber identity is concealed only to the casual observer. An investigating agent aided by the fused intelligence describe in the preceding paragraph will eventually determine attribution with a high degree of confidence. It will be a difficult process, but it is also endowed with guaranteed results. Everything is recorded in the cyber dimension. It's just a matter of gaining access to the cyber information, and the answer is awaiting. No one can disguise one's true identity and true source indefinitely. Given the required resources, that identity will be unveiled.

The process of attribution is indeed an arduous one, and therefore we cannot expect it to be a feasible solution for all cases of cyber attacks. Certain cyber intrusions certainly will not merit the effort of attribution analysis, specially when it comes to cases affecting cyber systems that are not part of the critical nodes on a enterprise network. However, for affected nodes that do merit the proper attention, due to their criticality in the network, forensic attribution is feasible.

Attribution, then, its an alternative regulated by the factors involved in any mitigation and recovery plan. When the required resources to apply forensic monitoring are available and merited, then attribution is possible, though not always achievable. On the technical side attribution requires very specific forensics procedures to collect enough data from local networks border routers, adjacent AS border routers, proxies and firewalls, DNS servers, authentication servers, and email servers. All these procedures require the

---

sources.
75 Giannelli, chapter 10, p. 107

intervention of professionally qualified and experienced cyber SMEs. On the analytical side, it requires the participation of professionally qualified cyber analysts capable of fusing technical and semiotic indicators.

The detection of the true source for a data stream that represents a threat to our personal or enterprise identity and operational security is best achieved only when we fuse in a synergistic manner the information proceeding from CYBINT, HUMINT, and SIGINT.

Semiotic analysis considers the study of all relationships occurring in the actions and methodology of the cyber adversary. It is impossible for a human being to execute an action void of personal and close community indicators. This is our personal, social, educational, ethnic and contemporary footprint. It is an unavoidable characteristic of our actions. Semiotic analysis allows us to detect, correlate, and ultimately create a cyber profiling of the characteristics of the adversary. If the importance of the affected target merits the utilization and associated costs of applying semiotic analysis, the results can be very productive, in conjunction with the technical forensic analysis.

Cyber anonymity and pseudonymity are only temporary disguises, so whether we are choosing one of this alternatives for protecting our personal or enterprise privacy, or for concealing our nefarious cyber activities, they are not permanent solutions.

Chapter 8. Cyber programming

Writing computer code is the art of organizing a series of sequential instructions designed to create an effect or an action executed by an electronic device designed to follow instructions. There is a considerable large segment of the global population that think computers are smarter than human beings. This is a fallacy created by the incognizance regarding cyber programming.

There is no intelligence on a computer system, in itself an electronic device that simply follows mathematical and logical instructions, executing exactly a set of instructions in order to produce an intended result. Cyber programming, therefore, requires a very precise sequence of operations written in accordance to the best possible style of clear expressions.

The forerunner of the modern electronic computer was conceived by the 19th century British mathematician and computing pioneer Charles Babbage. He designed the Analytical Engine, historically considered as the first fully-automatic calculating machine, but his design was only partially built before his death.[76] The Analytical Engine design included an arithmetic logic unit, control flow logic, and integrated memory.[77]

Ada Lovelace, a contemporary English mathematician, studied Babbage's Analytical Engine, and her notes on this subject contains sequences of operations designed to solve specific mathematical problems, namely, the first algorithm. Ada also envisioned the potential of the Analytical Engine to

---

76 http://www.sciencemuseum.org.uk/objects/computing_and_data_processing/1878-3.aspx
77 http://en.wikipedia.org/wiki/Analytical_engine

transcend mere mathematical operations. For her contributions Ada Lovelace is recognized as the world's first computer programmer.[78] There is an object-oriented high-level computer programming language, ADA, used in mission-critical applications, dedicated to honor Ada Lovelace.[79]

What Ada Lovelace accomplished in writing an orderly sequence of calculating tasks is the very essence of cyber programming, and it is not as easy as it may sound. Why? Because we are accustomed to expect that our instructions to accomplish a specific goal will be understood by another human being with the capabilities to fill in the gaps in our explanations. If you disagree with this statement you have not been living among humans.

Among English-speakers there is a severely abused speech crutch that repeatedly appears during the course of an explanation, punctuated by the ludicrous and annoying "you know what I mean", uttered whenever we fail to articulate all the necessary steps required by the intended explanation. When we are responsible for giving an explanation on a topic unknown to the interlocutor, it is unacceptable to interject this speech crutch and omit concepts or processes because we are unprepared to define them with accuracy and conciseness. We abuse this recourse because we expect the interlocutor to fill in the gaps.

This common deficiency in human communication illustrates the great challenge we encounter when attempting to communicate with a non-human, non-intelligent entity, namely, the computational device that requires specific, logical and concise set of instructions to perform a cyber task. In cyber programming the use of such speech crutch

---

78 http://www.computerhistory.org/babbage/adalovelace/
79 http://www.princeton.edu/~achaney/tmve/wiki100k/docs/Ada_%28programming_language%29.html

Cyber Reality

would lead nowhere, simply because a non-intelligent entity such as a computer cannot fill the gap created by a missing instruction in the series of logical statements required by the cyber processor. It simply has no cognition of the concept formulated in the mind of the programmer, since such processor is completely incapable of formulating detailed instructions inferred from an abstract concept.

The cyber programmer task is to meticulously define every single instruction in a logical, sequential, unambiguous and specific manner. The cyber processor is unable to infer any required step because it possess no intelligence at all. The cyber programmer has to decompose the abstract concept formulated in the human mind into the component minutiae integral to the abstract concept. The cyber processor is incapable of thinking; it can only calculate and apply the results associated with the programming instructions. Developing this set of orderly and coordinated instructions, the cyber code, constitutes perhaps one of the greatest test to our human capabilities in decomposing the abstract designs in our minds into cohesive and logical instructions written in a language germane to the operational capability of a cyber processor.

The language of conventional cyber processors is binary math, with the exception of QC systems. A conventional cyber processor operating on the traditional binary system can operate only on two possible values, namely, 0 or 1, thus the term "binary".[80] On the other hand, the QC system, following the conditions existing on a quantum system, is not restricted to either of these two only options. This is possible because of the superposition property in a quantum system, capable of holding a value of 0, or 1, or both values simultaneously.[81]

---

80 From the Latin term "binarius", formed of two components.
81 In the quantum realm the superposition property enables a particle to occupy all of its possible states simultaneously. See

87

Cyber Reality

Current studies on the nascent QC technology show a significant progress in operating with multiple qubits, thus exponentially increasing the possibilities of operating with multiple states for a given number of qubits.

Since computers are basically calculators, the language in binary systems process instructions to handle values via additions, subtraction, multiplications, divisions, and comparisons. The essential representation of computing instructions is expressed in binary language, used to represent any cyber object (numbers, characters, images, sounds), and sets of instructions in the computing device's native language. However, the codification in binary language may become overwhelming, and we use alternate representations to facilitate reading and storage in cyber memory. Thus, we refer to this alternate nomenclature as octal notation, and hexadecimal notation. These two alternate nomenclature simplify the binary nomenclature. For instance the letter A requires the binary representation 01000001, but in octal nomenclature this value is reduced to 101, and in hexadecimal nomenclature is simplified as 41.

As you can surmise, writing cyber code in native language (binary) becomes an overwhelming task, especially if the cyber code contains millions of line of code present in current cyber programs. Most current software is written in high-level languages, where we use an abstraction methodology to express cyber instructions while avoiding to write them directly in native binary language. High-level languages abstract the direct communication with the specifics of the microprocessor with the assistance of a multitude of large sets of programming libraries.

The evolution of programming languages can be succinctly,

---

http://www.princeton.edu/~achaney/tmve/wiki100k/docs/Quantum_superposition.html

Cyber Reality

while not comprehensibly described, as the progression from native language to high-level language. First generation languages (1GL) encompass the very early (1940s) set of instructions consisting entirely of binary representations, also known as machine language. 1GL does not need compilers or assemblers, because 1GL are directly executed by the computer's Central Processing Unit (CPU).

Second generation languages (2GL) are based on the use of symbolic names instead of purely binary language. Introduced in the 1950s, 2GL is also known as assembly language since it requires an assembler to translate the cyber code into machine language. With the arrival of third generation languages (3GL), also introduced in the 1950s, we became able of reaching a higher level of abstraction, by using words and commands instead of simply symbols and numbers, as in the case of 2GL and 1GL, respectively. The 3GL or "high-level languages" include the well-known C, C++, Java, and Javascript programming languages, among others. 3GL require the intervention of a compiler, in order to interpret or translate the high level instructions into machine language. Other programming language generations exist, but they are designed for more specific, not general purpose environments.[82]

So the task in front of the cyber programming is to use a programming language that can be processed by a cyber processor operating in a numerical binary format. This is known as the native machine language. How then can a human being communicate with a cyber processor that speaks a very peculiar native machine language? The answer to this challenge is the oldest programming language, known as assembly language. It consists of a

---

[82] 4GL and 5GL, introduced in the 1970s and 1990s, respectively. For additional information on the topic of programming code you may consult http://landofcode.com/programming-intro/writing-computer-programs.php

series of very concise instructions, where each one of them corresponds to a singular machine language instruction.[83]

How do you communicate with a computer, and provide it with detailed and precise instructions of what you want done by it? Let's remember, we are attempting to communicate with a calculator, void of intelligence, incapable of deducting reasoning. You simply cannot expect this calculator to "know what you mean". All that it knows is what you tell it, and what you tell it must come in the form of values that can be processed by the computer's processing unit, in machine language. Do you remember what we said about Ada Lovelace and her contribution to computer programming, and why she is honored and remembered as the "first programmer"? If you said: "algorithm" then you are correct!

We communicate with computers via algorithms, a series of coordinated and concise steps for executing a specific task. Thus, algorithms are the basis of our methodology in communicating with a computer and enable it to arrive at the completion of a task, the solution for a problem. And how do we document the series of instructions, the algorithms? We create a source code, the actual text used to express in written form all the instructions for our computer program. However, since we do not communicate in machine language, we need a translator between our source code and the computer's processor.

Our next step into the creation of a computer program is to employ a compiler, the translator needed between us and the cyber processor. The compiler is the software tool that translates source code into the data format understood by the processor. A compiler, therefore, is the translator that transforms our source code into object code, and this object code eventually becomes an executable program. This

---

83 Kip Irvine, Assembly Language for Intel-based Computers, Fifth ed. (New Jersey: Pearson Prentice Hall, 2007), 1-4

Cyber Reality

statement is of course a very cursory review of the process of computer programming, intending to simply illustrate the entire general process while omitting many technical details.

As a gifted thinker and refreshingly candid observer[84] already stated it, computers can only interact with human beings at the highest level of data integrity. If the human programmer is unable to express a set of instructions in a clear, concise and coordinated manner, the results will not be the desired ones. A computer does not pretend to understand an ambiguous instruction; it will simply rejected it. A computer does not pretend to comprehend an incongruous statement simply to flatter the human programmer, hoping to be rewarded with a career advancement for covering mistakes generated by the human programmer. A computer is an extremely and brutally "honest" partner. A computer will simply respond to any given instruction properly expressed in computer language. If the instruction is correctly expressed, the result will be the expected result. If the instruction is flawed, incorrectly stated, illogically conceived, and systematically anachronistic, the result will not match the desired result. Nothing test the capacity to deconstruct a procedure in its orderly components as the task of computer programming.

This concept of communicating with a computer in a precise, concise and systematically correct manner may appear foreign to the GUI[85] generation, so accustomed to think that all it takes is simply to click a little icon or touch it directly on the screen of the device with which we want to interact. Perhaps the ubiquitous GUI has become the worst invention in the cyber realm, robbing us of our ability to communicate with a computer in a logical and intentional manner. The democratization of computer use has made possible for the

---

84 James P. Hogan, Mind Matters: Exploring the World of Artificial Intelligence (New York: Del Ray, 1997)
85 Graphical User Interface

masses to benefit from the use of computing devices, but it has also plunged us into the utmost ignorance on the art and science of communicating with a computing device.

How many of the readers are capable of communicating with a computing device and perform computing activities from the command line? How many readers do even know what the command line is, and how to access it? How many readers would feel powerless if they were deprived of the use of the mouse? How many readers are capable of launching a computing application by typing an instruction string from the command line? The ubiquitous icons on our current cyber devices GUI are simply the middle man between the user and the computer's operating system.[86] GUIs introduce a degree of separation between the computing device and the user, and GUIs were never intended as part of the original interactive design between the user and the system. GUIs are an afterthought, a concession imposed by the mediocre and demeaning mindset that always assume humans are incapable of interacting with a cyber system by issuing commands from the CLI.[87]

It has been erroneously argued that the GUI is a necessity for the sake of mitigating the so-called steep learning curve of using the CLI. This is a fallacy, sponsored by marketing schemata, directed at promoting the perception that the use of CLI represents a more difficult manner of interaction with a cyber system. By promoting the perception that the GUI is "easier" than the CLI, the marketing schemata expects an increase on the sales of devices equipped with GUIs. The truth of the matter is that GUIs limit the use and depth of interaction between the user and the cyber device. An accurate comparison of the scope of communication between GUI and CLI can be accurately expressed as the

---

86 OS hereafter
87 Command Line Interface

following: The GUI is to cyber communication as the act of pointing is in human conversation. Would you reduce the extent and scope of your conversational skills to the simple act of pointing, and perhaps adding grunting for emphasis?

There are several orders of magnitude of difference between what a user can communicate to a cyber device when using either GUI or CLI. Do you have any idea of the depth and scope of communication you can have with your cyber system when using CLI? Do you have any idea how limited you are in communicating via GUI? Imagine attending a social event where you can interact with individuals possessing a great depth of information of various topics, but you can only point to a chart with simplified symbols in order to communicate with them? This is exactly what you are doing when interacting with a cyber device only via the use of the GUI. The good news is that you are not limited by using the GUI only. The CLI is always available for your use, ready to enable you in transitioning to a much higher and deeper level of communication with your cyber device.

The CLI not only will open an amazing door into the depth of the data information in your cyber system, but it will also usher you into the amazing degree of control you can exercise on the behavior of your cyber device, away from the shallowness and superficiality provided by the GUI. If all that you aspire is ease of use and shallow answers, stay with your beloved GUI. If you want to become the master of your cyber system, and exercise complete control on your cyber system, then step into the CLI sphere of power.

Allow me to give just one example of the debilitating effects of limiting yourself to the use of the GUI, as opposed to the empowering effect of the use of the CLI. It is a common occurrence that anyone using a web browser will inevitable land on a persistent site, characterized by the on-the-fly modified behavior on the visiting browser. When arriving at such sites the user discover that the browser's back button or the close button are disabled, and the user is not allowed to leave the visited site enabled with persistent behavior. For those enslaved by the GUI mistress there is no solution but to reboot the system, since even the Task Manager (for Windows users) is unable to terminate the intrusive browser session. Those of us who are practitioners of CLI do not have that problem, and our absolute control over our browser, when visiting a persistent site, allows us to expeditiously terminate, without any delays, the intrusive browser session. Just a simply CLI command, and we are back in control, without rebooting our system, ready to continue our browsing at will, assuming this system is using a flavor of the Linux OS.

Why dedicating several paragraphs to the discussion of something as simple as the use of the GUI? Because behind the marketing schema of the GUI promotion lies the CYRE of control and power over the information capabilities of your cyber system. This book is dedicated to the topic of cyber reality, a reality that remains hidden from the majority of users, because they have been misled into perceiving a pseudo CYRE that has been artificially divested of its depth and richness. The true dimension, power and scope of CYRE has been reduced to a shallow shell, sacrificed in the altar of the marketing schemata, reduced and diluted for the sake of making CYRE more palatable to the masses that prefer mediocrity over excellence. All that the GUI requires is ease of use and oversimplification, while the CLI requires a degree of effort and commitment. Choosing the latter, however, ushers us into a realm of richer and deeper interaction with our cyber systems. This choice divides us into the two groups of cyber users we know today: the cyber-know and the cyber know-not.[88]

The fallacy promotes by the GUI promoters hides the true

---

88 See chapter 10 ahead

nature of the GUI. The major problem with GUI front ends is their role, a middle man between the OS and the user. GUIs frequently lack the versatility, granularity and control offered by CLIs. GUIs offer limited options and lack the command concatenation capabilities enabled by CLIs, via the use of pipes and switches, allowing the user to perform highly specialized tasks unavailable in GUIs.

This is one reason that experienced users of Unix-like operating systems often prefer command line programs over their GUI counterparts. Another reason is that command line programs in some cases can be quicker and/or more reliable than their GUI front ends. In addition, command line programs are often the only available choice for repairing and restoring damaged systems limited to operate with minimal functionality.

Allow me to give you an example of the power of CLI in combining commands into a single line, with very accurate and expeditious results:

```
ls /home/user | tee my_directories.txt
```

With the preceding CLI instruction I can perform several actions with a single command. This CLI command will list the contents on /home/user and this list will be sent both to the standard screen output and it will also be written to a file with the name "my_directories.txt". I challenge any GUI user to replicate the same results by clicking a single icon in the GUI.

Let's now revisit the concept of control over one's cyber system with a GUI user, by reiterating the annoying and potentially dangerous experience of getting caught into visiting a web site configured with a persistent page that disables the back and the close buttons in the browser's menu. While held captive in a particular page, the system

Cyber Reality

may be subject to the collection of the user's personal habits and even the user's personal data. Panicking and repeatedly clicking on the back button will not provide any escape, because the HTML code is written as to hold the browser captive, against the will of the affected user.

Let's us now contrast this powerless and potentially dangerous scenario with the CLI capabilities. This author enjoys a Linux-only environment at home, and maintains complete control on the behavior of his Linux cyber systems, and with the execution of a single CLI instruction can instantaneously escape any web site attempting to hijack the control of his browser. This complete control on one's cyber system(s) is the essence of CYRE. A cyber system is not a master. It rather exists as a working horse dedicated to the service of his owner, and under the complete control of his owner. It is unnatural to witness so many cyber systems dictating anomalous behaviors over the will of their owner, and even sadder, when such owners act in complete subjection and powerlessness to such behavior. In the purest expression of CYRE, man is the master of the cyber device, not the other way around.

A computer does not pretend to understand an ambiguous instruction; it will simply rejected it. A computer does not pretend to comprehend an incongruous statement simply to flatter the human programmer, hoping to be rewarded with a career advancement for covering mistakes generated by the human programmer. A computer is an extremely and brutally "honest" partner.

Is there a price to achieve this degree of control and

Cyber Reality

performance from our cyber systems? Of course there is a price, given the axiom that in this our universe there is no such thing as a free lunch. If you want to achieve a level of excellence in using cyber systems, and controlling them so they perform to their optimum capacity, we have to learn how to efficiently communicate to them our instructions. The GUI is a crutch for those who are contempt with a superficial performance and a quick answer. For those who seek a degree of mastery in commanding this cyber force, then we have to enter into the power zone of the CLI. And what do we do when the mouse becomes frozen and the icons deactivated? What do we do when we require a more comprehensive and more reliable data flow from our cyber system? We step into the CLI zone. Furthermore, the CLI becomes the only available means of communication with a cyber system when repairing and restoring tasks are required to renew a system with degraded functionality.[89]

Cyber programming is the discipline that ushers us into the cyber dimension, is the language that allows humans to interact with the cyber devices we have designed. It is very important we do not lose sight of this reality. We humans designed these cyber devices we deployed around the world, and as designers we should retain our role as masters. However, the current relationship which the majority of the members of the global village maintain seems to rapidly deteriorate into a relationship of servitude. The general population seems to assume that cyber devices know better than us humans. We seem to accept that we are no longer in control of the designed cyber devices; we tend to feel quite content with surrendering control to the cyber devices.

And why is that? Because we have forgotten, or because we never knew, that cyber devices are tools we have

---

[89] The Linux Information Project:
http://www.linfo.org/command_line_program.html

created, and therefore we command. But how could we command them if we seem not to know how to communicate with them? How many times have we heard the comments of an older family member stating how amazing is the fact that very young people seem to have an innate control over the cyber devices they use? And what about the visible contrast between these young people and the older members in the same family, who tend to feel intimidated and almost powerless when attempting to use the same cyber devices?

The difference resides in the attitude. The younger family members understand the cyber devices are tools at the service of human beings, and therefore these young people take control of them by trial and error. It's the attitude leading younger people to say: I will take control of this cyber device and make it do what I want it to do. This is the essence of the role of cyber devices; to do what we human beings want them to do. Cyber devices should be at our service.

Computers are not smart. They simply execute instructions provided by human beings. Conversely, computers do not make mistakes. The mistakes arise from faulty instructions. The more exquisite, orderly, methodical, organized these instructions are, the more amazing the results. We humans control the cyber age, not the other way around.

Chapter 9. Abridged History of Computing

The cyber dimension is an unprecedented reality in history, a unique product of the twentieth century. Computing, however, is as old as mankind. Computing is a cognate of calculating, a methodology seeking automation and simplification in achieving results to recurrent and repetitive tasks in society, extending to a variety of human disciplines, including science and industry.

A computer, in its fundamental characterization, is a calculator, designed to automate tasks involving the resolution of equations via calculations. We have documented cases of calculating devices in use as early as 2700 BC, and all these early devices, and their corresponding improvements and variants, were depending on an automated series of parameters arranged in a certain order, and operated in a certain sequence, in order to achieve a desired result.

Some of the most significant landmarks pertaining to computing prior to the 20th century can be summarized by referencing the development and use of earlier calculator-computers. There are records indicating the use of abaci in the BC period of ancient Mesopotamia, Egypt, Persia, Greece, and Rome. Many other cultures became users of this calculating device in the AD period, and several maintain the use of abaci even as of today, including Chinese and Japanese users. The former refer to this calculator as "Suanpan", while the latter refer to it as "soroban".

The table below marks some of the landmarks in history when a calculating device of special significance was adopted for computational tasks.

Cyber Reality

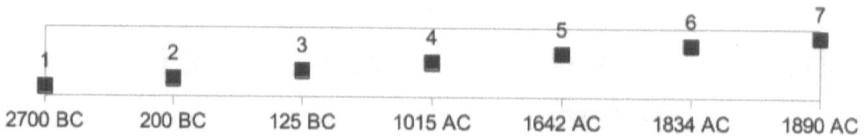

1 = Sumerian abacus
2 = Chinese abacus
3 = The Antikythera System (first astronomical analog computer)
4 = The Equatorium (astronomical analog computer)
5 = The Machine Arithmetique (Pascal's analog mechanical computer)
6 = The Analythical Engine (Babbage's firts mechanical computer)
7 = The Tabulating Machine (a census calculator, precursor to IBM)

Coming into the 20th century we can see the arrival, progression, and ascending evolution of the calculator-computer entering the electronic age and the rapid progress in computational capabilities.

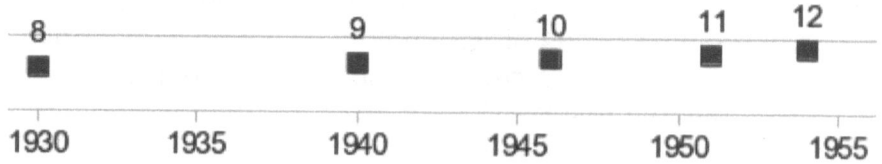

8 = The Difference Engine (partly electronic equation calculator)
9 = The Complex Number Calculator (Bell Labs)
10 = ENIAC, the first electronic, programmable computer
11 = Whirlwind, the first interactive computer
12 = FORTRAN, the first high-level programming language.

It is quite acceptable to state that the history of modern electronic PC began in 1947 with the invention of the transistor by scientists at Bell Telephone Laboratories. The miniaturization of electronic circuits made possible by the transistor became the key factor in the development of small, reliable, and affordable personal desktop computers. The chart below shows a time-line for some of the important landmarks leading to the ubiquitous PC in modern life.

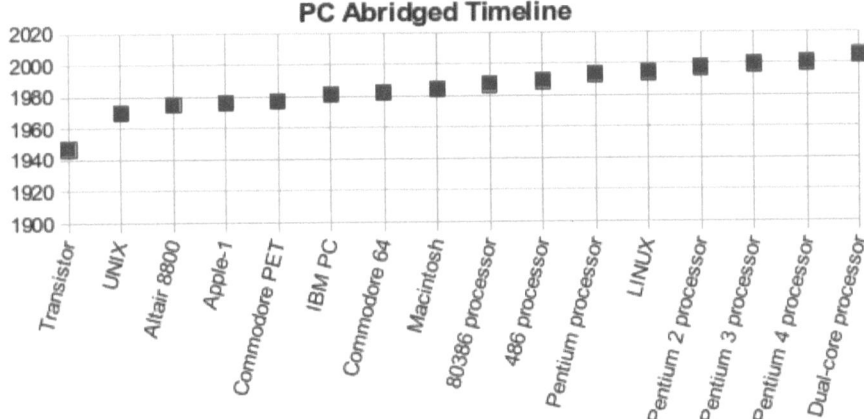

And finally, and most recently, in our own contemporary history during the two decades of the 21st century, we have witnessed the accelerated progression of the High Performance Computing (HPC) systems, and the Quantum Computing (QC) systems. The theoretical foundations of QC were established during the 1980s, and the emergence of HPC systems occurred during the 1990s.

The HPC is a computing cluster platform operating under quite different parameters than the traditional PC operating at the enterprise or residential level. An HPC system is designed to perform large-scale mathematical calculations on complex projects, especially in scientific and engineering fields. The cluster computing design is based on parallel computing, designed to deconstruct a large and complex task into numerous smaller tasks that can be performed simultaneously, thus allowing us to achieve completion at a faster rate.

The cluster design of parallel computing on HPC systems requires new architectural models for memory management, operating systems, networking, interconnects, energy and heat dissipation to operate the numerous multi-core processors reaching into the seven digits[90] realm.

When we deal with large-scale mathematical calculations it is imperative to optimize the speed of floating-point operations, central to the design of the HPC system. Floating-point arithmetic is essential in all computing systems, from PCs to HPCs, since is the foundation for the measurement of computing performance. Floating-point calculation rate is commonly known as FLOPS.[91] The importance of floating-point operations in computational design is masterfully described on a paper published in 1991.[92]

A regular PC with a single-core processor operating at 2.5 GHz has a theoretical performance of 10 GFLOPS.[93] Furthermore, even with the arrival of PC with dual and quad core processors, they still remain at the GFLOPS level. By comparison, and two orders of magnitude above, the current leader HPC system since November 2013 has a performance of 33.8 PFLOPS.[94]

We can visualize the substantial difference between traditional PC performance and HPC performance by referring to the following chart, showing the order of magnitude different scale between the performance level of

---

[90] The current number 1 position HPC system in the November 2013 list uses 3,120,000 cores.
[91] Floating-point operations per second.
[92] David Goldberg, What Every Computer Scientist Should Know About Floating-Point Arithmetic, March, 1991 issue of Computing Surveys. Copyright 1991, Association for Computing Machinery, Inc.
[93] 10 billion FLOPS
[94] 33.8 quadrillion FLOPS

regular PCs and HPC systems.

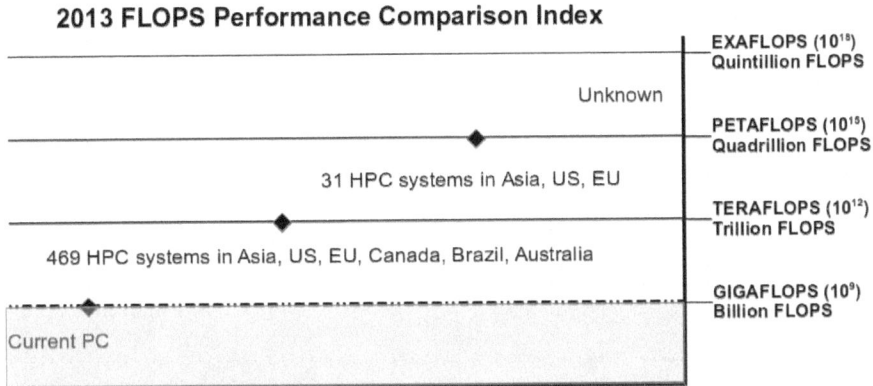

However, of the 31 HPC systems operating at the PFLOP level, there are only 4 of them operating in the double-digit category, as illustrated in the following chart showing the current November 2013 Top 10 HPC systems. Only Asia and the US belong in this double-digit category. The single-digit category encompasses the EU and the US.

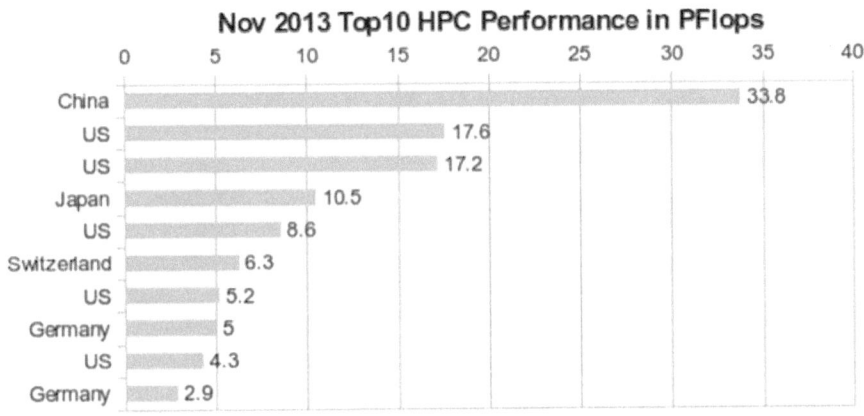

When we look at the global distribution of the current

Top5000 HPC systems, the US has the vast majority, followed by Asia, and Europe. This distribution schema explains the reason why the leadership in HPC systems fluctuates between the members of these three groups, as seen in the chart below.

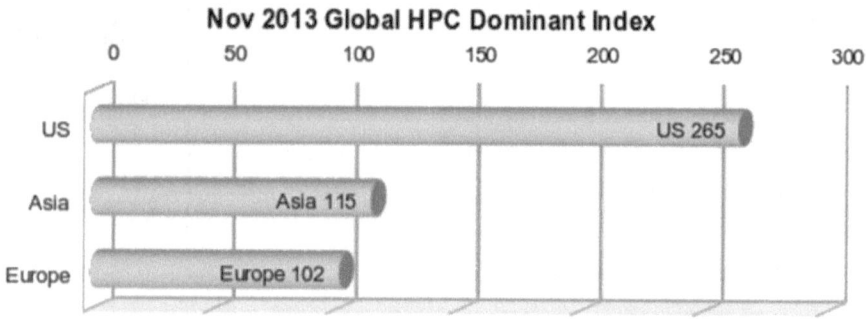

The majority of HPC systems in the Asian continent are in China, followed by Japan. The distribution of HPC systems in Europe is basically shared by UK, France and Germany, as shown in the two following charts.

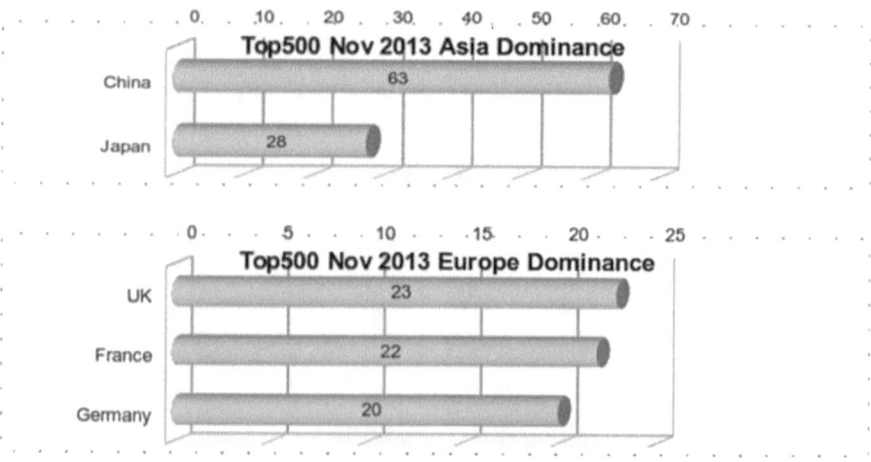

The foundations for the three computational systems we are developing and using today were established during the last three decades of the 20th century, namely, the PC systems, the Quantum Computing (QC) systems, and the High Performance Computing (HPC). The chronological progression is illustrated in the following chart.

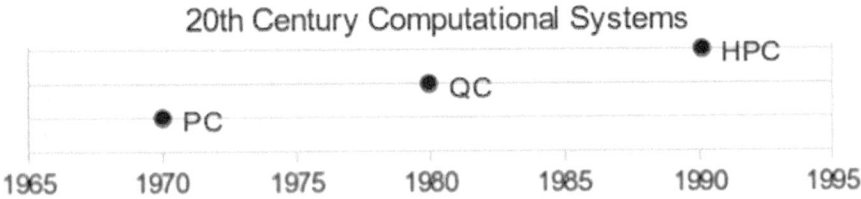

The brief history of the rapidly ascending computational performance above the GIGAPLOPS began at the very end of the 20th century, when the Intel's ASCI Red became the first HPC in the world to break the TFLOPS barrier, achieving a Linpack performance of 1.068 TFLOPS in June 1997. This HPC retained the Top500 No. 1 position until June 2000, with an upgraded Linpack performance of 2.379 TFLOPS, and remained in active service for the US Department of Energy (DoE) until 2005. ASCI Red marked the landmark of the beginning of U.S. dominance in the production and employment of HPC systems.[95]

The Top500 No. 1 position remained in the US from 1997 until November 2001. The Top500 leadership position went to Japan in June 2002, where it remained until June 2004. In November 2004 the Top500 leadership position returned to the US, who retained the leadership at the TFLOPS level until November 2008.

Then, in June 2008 the US once again set another important landmark in the history of HPC systems, by breaking the barrier into the PETAFLOPS performance level. The US retained the leadership at the PFLOPS level until June 2010.

---

95 http://www.top500.org/featured/systems/asci-red-sandia-national-laboratory/

Cyber Reality

The next Top500 list placed the Chinese Tianhe-1A HPC system in the leadership for the November 2010 list.

In June 2011 the leadership position went back to Japan, where it remained until November 2011. However, in June 2012 the US regained the leadership position, and retained it until November 2012. Then China made a comeback to recover the leadership position in June 2013, and maintained it in November 2013. The chart below illustrates the abridged history of leadership on HPC systems.[96]

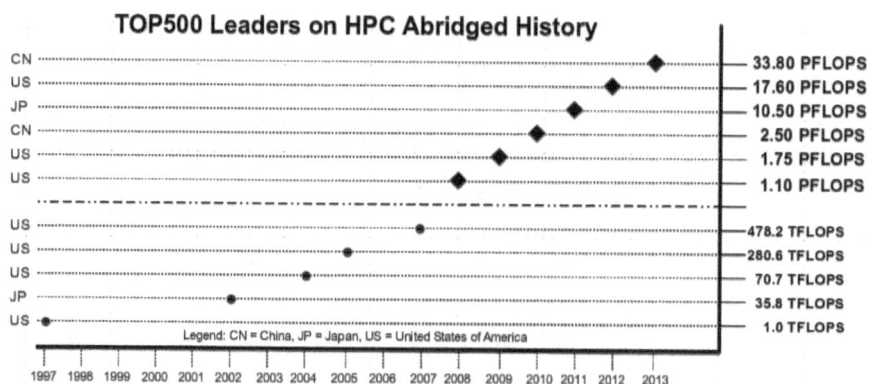

This chart highlights the fact that the HPC leadership fluctuates only between three countries, and that the progression trend has decelerated since we broke into the petaflop scale. This may signal the point in our performance progression where the parallelism in the HPC processing is becoming affected by the limitation expressed in Amdahl's law.[97] There is a point when parallelized applications reach a scalability limit, and the increase on additional processing cores will no longer increase performance.[98]

---

96 The data for this chart was compiled in November 2013 directly from the site http://www.top500.org/lists/
97 Gene Amdahl, the American computer architect who stated the fundamental limitation of parallel computing.
98 Jeff Layton, Why Isn't Your Application Scaling?, http://www.admin-magazine.com/HPC/Articles/Failure-to-Scale.

Amdahl's law addresses the fact that the theoretical speed increase of an application operating with multiple processes will inevitably reach a limit set by the serial performance portion of the same application. There is an important difference between Amdahl's Law and real-life HPC parallelism; we do not have an infinitely fast network, with zero latency and infinite bandwidth. In real life we do have to cope with latency greater than zero and limited bandwidth, and data requires a certain amount of time to be retrieved, to be written, and to travel through a network less than ideal.

So, what's the road ahead to mitigate this limitations? A feasible solution may reside in revising (and rewriting when necessary) the optimization factor of both the parallel and the serial portion of the code on the application being processed. The latter may represent the most serious challenge, since it will entail to deal with the I/O portion. When the application starts it requires to input data, and during operations requires to write data and store results at the end of operations. This is due to the fact that typical I/O[99] operations take place in a single process, in order to avoid collisions. Thus, very careful programming is required to allow multiple processes writing to the same file, in an optimized manner, while avoiding collisions.

Careful programming in itself is not enough; we have to factor network latency and bandwidth limitations, that together may impose periods of waiting, affecting the serial coordination of I/O processes.

The limitations we are experiencing in our quest for the next step, that of transcending the petascale limit and reaching into the exascale next order of magnitude in computing performance, may also serve to lead us into recovering our sobriety. There are many utopians in our present society,

---

99 Input/Output

Cyber Reality

who lured by their unfounded enthusiasm in the prowess of electronic man-made computing, are envisioning an age when computers will substitute human beings. This unrealistic expectation is also part of the distorted CYRE adopted by many in our contemporary history.

To put it in very simple terms, computers are calculators guided by instructions designed to optimize the performance of a particular task, and they are completely dependent on humans. Humans, on the other hand, profit from the use of this calculators, but they are not dependent on them. Computers are a tool, but humans are superior beings.

A recent article[100] describes how an experiment conducted by Japanese and German researches illustrates the superiority of the human brain over computers. They attempted to simulate brain activity by employing an HPC system operating with over 80,000 processors. It took them 40 minutes to replicate just one percent of one second of human brain activity. The HPC system used for this simulation requires almost 10 MW of power to operate.

At the closing of 2013 an article on a computer magazine[101] brought us a sober reminder that on the HPC race for superiority there are significant challenges to overcome, and a sound strategy is necessary to maintain or advance our HPC superiority status. Even though Hewlett-Packard, IBM and Intel collectively have the majority (over 70%) of the petascale HPC systems on the current Top500 list, the US faces hard competition from the European Union (EU), China and Japan in the quest toward the exascale level. While the EU already set a budget for an exascale system

---

[100] George Dvorsky, online article by the Instititue for Ethics and Emerging Technologies, August 25, 2013, http://ieet.org/index.php/IEET/print/8077
[101] Patrick Thibodeau, U.S. Falling Behind in Exascale Race, ComputerWorld, December 16, 2013, http://www.computerworld-digital.com/computerworld/20131216?pg=7#pg7

scheduled for 2020, and China and Japan are also pursuing the same goal and schedule, the US has made no public announcement with regard to either a budget or a specific schedule for reaching the exascale level. China is committed to a very significant investment in the HPC arena, and Japan has a definite plan they expect to fulfill in reaching the exascale level.

At the moment there are no clear indicators that USG leadership is concentrating on HPC R&D as a priority. We certainly hope this current trend mutates into a more aggressive and substantial commitment to pursue the exascale goal, so as to maintain and enhanced US scientific, technological, strategic and tactical advantage.

Update information. In June 2014 this author attended the HPC conference in Leipzig, where a trend was unveiled: the historical growth trend in HPC performance is exhibiting a decrease. In order to better understand this performance decrease is prudent to consider a retroactive summary of the TOP500 List, the mechanism used since 1993 to maintain a historical record of the HPC performance at a global scale.

The following table shows the progression in HPC performance, from the inception of the TOP500 List to the present. The data is taken from the records maintained by the official TOP500 List.[102]

The scientific nomenclature for measuring computational performance is expressed in FLOPS. This acronym represents FLoating-point Operations Per Second. The scale in the following table includes GFLOPS (GigaFLOPS) as billions of computational calculations per second, TFLOPS (TeraFLOPS) as trillion of computational calculations per second, PFLOPS (PetaFLOPS) as quadrillion computational calculations per second, and the

---
[102] http://www.top500.org/

next order of magnitude higher scale of EFLOPS (ExaFLOPS), as quintillion computational calculations per second.

The inclusion of EFLOPS in this table is intended to illustrate the fact that many auspicious, and perhaps overly optimistic, announcements were made prior to June 2014, anticipating that HPC systems would transition into the EFLOPS scale by 2016-2018. However, after the consensus reached at the Leipzig 2014 ISC meeting, the HPC community knows that there is a documented slowdown in the performance trend. Thus, the same voices that were anticipating HPC at the Exascale level by 2016-2018 are now mute.

Some news media spoke of exascale online by the end of the decade,[103] while other more optimistic announcements spoke of exascale systems by 2018, operating under the optimum power consumption mark of 20 megawatts.[104] Other estimates maintain the more generic prediction of exascale systems by 2020, but with a consumption power of less than 30 megawatts, above the optimum mark.[105] From this short review it is apparent that the more recent

---

103 Matthew Knight, CNN, Faster than 50 million laptops -- the race to go exascale, March 30, 2012, http://www.cnn.com/2012/03/29/tech/super-computer-exa-flop/
104 Intel European Exascale Labs Report 2011, http://www.exascale-computing.eu/wp-content/uploads/2012/02/Exascale_Onepager.pdf
105 Patrick Thibodeau, Computerworld, Why the U.S. may lose the race to exascale, Nov 22, 2013, http://www.computerworld.com/article/2486248/high-performance-computing/why-the-u-s--may-lose-the-race-to-exascale.html

predictions tend to be more conservative than earlier ones.

From the Asian sector we have heard announcements from Japan and China. The former is aiming to deploy an exascale HPC system by 2020, while the latter has announced plans to reach the same goal before 2020.[106] India appears to be the most optimistic of all, by announcing their plans of deploying an exaflop HPC system by 2017.[107] Obviously, this is not a very judicious expectation, considering that India does not have any leading PFLOPS HPC system in the first top 10 in the official TOP500 List, dominated by China, US, Japan, and some European nations. India is still operating at the lower TFLOPS level, with one HPC system in the 52nd position of the current TOP500 List.

Considering that China has also announced a time frame for building an exascale HPC system between 2016-2020,[108] this author personally contacted the NUDT official representative after attending the plenary session at the HPC conference in June 2014. The announcement of the performance plateau was announced during this session. This author asked the NUDT representative if China was still maintaining this optimistic goal, but this representative declined to answer the question.[109]

---

106 Computation Institute, Rick Stevens on the Race to Exascale, 23 May 2013, https://www.ci.uchicago.edu/blog/rick-stevens-race-exascale.
107 Robert Gelber, HPCWire, India Sets Sights on Exascale for 2017, September 18, 2012, http://www.hpcwire.com/2012/09/18/india_sets_sights_on_exascale_for_2017/
108 Richard Yonck, World Future Society, Exascale Supercomputers: The Next Frontier, March 22, 2011, http://www.wfs.org/content/exascale-supercomputers-next-frontier.
109 Event took place at the Exhibition Hall of the ISC'14 conference in Leipzig, Germany. The National University of Defense Technology (NUDT) is the organization responsible for operating Tianhe-2, the current HPC lead system in the TOP500 List June 2014.

As stated before, the earlier the predictions for an exascale HPC systems, the more optimistic, with many earlier publications pointing to 2018 as the dawn of the exascale HPC system.[110] Yet, after the HPC 2014 conference, the estimates are growing more conservative, and there are some voices stating that even 2020 is an unachievable goal. A reputable expert in HPC performance has issued the challenge that 2020 will not see the dawn of the exascale HPC system, and has placed his reputation and personal money to support his view.[111] Why would somebody have such a dissimilar views with the rest of the optimistic predictions we have briefly outlined in this update section?

Breaking the barrier into the exascale level will require more than simply applying a scalability model. It is not just a matter of adding more cores into the design of our current HPC systems. Processing power is just one of many challenges, including power consumption, heat dissipation, memory management, and interconnect optimization among others.

Whenever we may reach the goal of transitioning into a quintillion computations per second we will also be dealing with a substantial growth in picojoules, the electrical-mechanical-thermal unit. While theoretically possible, an exascale system within the current parameters of available technology would exceed a dedicated 100MW of power to sustain operations of such system, far above the optimum goal of 20MW range.[112] By comparison, a 100MW of power

---

[110] HPC Advisory Council, Toward Exascale computing, 2010, http://www.hpcadvisorycouncil.com/pdf/Toward_Exascale_computing.pdf

[111] Joel Hruska, ExtremeTech, Supercomputing director bets $2,000 that we won't have exascale computing by 2020, May 17, 2013, http://www.extremetech.com/computing/155941-supercomputing-director-bets-2000-that-we-wont-have-exascale-computing-by-2020.

[112] Ibid

consumption is equivalent to the power consumption of 80,000 US households.[113]

A reputable study on the challenges presented by an exascale system highlights several of the issues at stake. For instance, as power and cooling constraints limit increases with microprocessor clock speeds, the desired solution sought by HPC developers resides in increasing on-chip parallelism to improve performance.[114] A corollary to this issue resides in the energy cost of moving data off-chip, characterized as the biggest change in energy cost. This is required for moving data to different levels of system memory relative to the cost of a floating point operation.[115]

The cooling and power distribution challenge is quite evident. If we consider even the least expensive power in the US at the cost of less than 5 cents per KWH, an HPC system would operate at a cost of USD 1 million per megawatt per year. This estimate will assist the reader in understanding why the HPC industry is adopting the limit of 20MW for a viable HPC system design.[116]

Another consideration in the exascale challenges is the power consumption required by current memory technology such as DDR3, at its current bandwidth/flop ratio of 1 byte/flop. The power consumption of a system with only 0.2 bytes/flop of memory bandwidth would exceed 70MW, thus not a desirable design. On the other hand, maintaining the design under the 20MW goal would force the memory

---

113 Chris Nuttall, The Financial Times, Supercomputers: Battle of the speed machines, July 9, 2013, http://www.ft.com/cms/s/2/24dc83d6-e40e-11e2-b35b-00144feabdc0.html#axzz3Bq0Syr3x
114 Shalf, Dosanjh, and Morrison, LLNL, SNL and LANL, Exascale Computing Technology Challenges, 2011, https://www.nersc.gov/assets/NERSC-Staff-Publications/2010/ShalfVecpar2010.pdf
115 Ibid
116 Ibid

system to have less than 0.02 bytes/flop. This design would severely limit the number of applications to efficiently run in such system.[117] The challenges the HPC industry has to overcome in pursuing an exascale system are not trivial.

The plateau in HPC performance announced at the ISC'14 conference in Leipzig, however, is not an isolated event. There are two previous major precedents that become evident when we explore the historical data on HPC performance progression on the number 1 position in the TOP500 List during the last 14 years. The data on HPC performance since 2000 indicates that we have experienced three minor and three major occurrences of performance plateau. Among the former category we have an occurrence between June 2001 and November 2001 (at 7.2 TFLOPS), then between November 2008 and June 2009 (at 1.1 PFLOPS), and finally between November 2009 and June 2010 (at 1.8 PFLOPS).

The three major occurrences are considered more significant because the duration of the performance plateau extended over a longer period of time, when compared with the minor occurrences of the performance plateau. The following graphic illustrates the three major occurrences of the performance plateau, as measured in relation to the number 1 position in the TOP500 List between 2000 and 2014.

---

117 Ibid

Cyber Reality

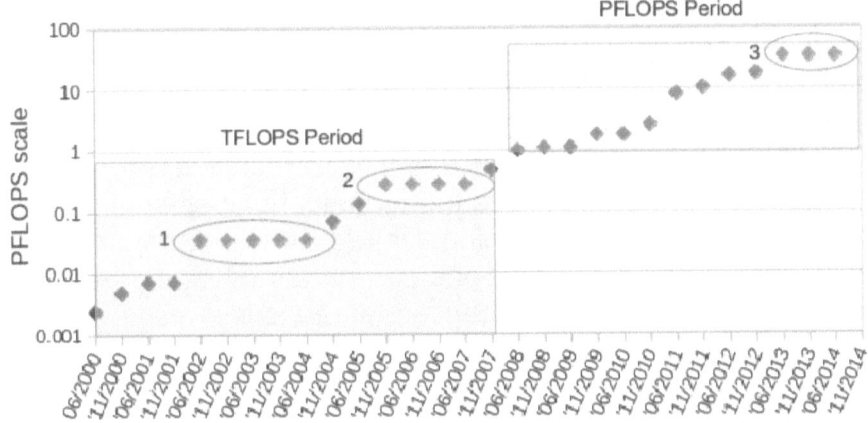

The three major occurrences of performance plateau are indicated by the three ellipses, labeled 1, 2, and 3. Two of them occurred at the TFLOPS level, and the third at the PFLOPS level. The first occurred between between June 2002 and June 2004, when the HPC performance remained at a sustained rate of 35.9 TFLOPS. The second occurred between November 2005 and June 2007, with a sustained rate of 280.6 TFLOPS for that period. The most current performance plateau began in June 2013 and continues at the time of writing of this updated section,[118] at a sustained rate of 33.9 PFLOPS.

Whatever might be the interpretation of this last performance plateau, the main factor to consider is the vast scope of this most recent plateau, affecting 9 out of the Top 10 HPC systems. This extended scope is indeed unprecedented.

On occurrence number 1 the performance plateau affected only the the leader HPC system in the top 10. On Occurrence number 2 the performance plateau affected the leader HPC system, and partially the second system in the top 10. However, in the current performance plateau period,

---

118 This statement was written on 7 September 2014.

115

50% of the top 10 HPC systems were affected in November 2013, but 90% of the top 10 were affected in June 2014. This is the factor that renders this most recent performance plateau as unprecedented and uniquely significant. Since the HPC performance rate is a highly competitive arena, and the affected systems involve the technical and scientific prestige and prowess of China, United States, Japan, Switzerland, and Germany, one can hardly advance the hypothesis that this performance plateau is just an unintended coincidence.

There is one concrete factor that may contribute to a hypothesis with enough merits to be considered feasible: the enormous funding resources required to break the current scientific and technological barriers we have cursorily presented in this chapter. The value of this hypothesis resides in its elegant simplicity. What may cause all the nations represented in the top 10 HPC systems to reach a simultaneous performance plateau? Without the required funding the players cannot achieve any scientific and technologically revolutionary results necessary to break the performance barriers. This task becomes even more difficult when attempting to break the exascale barrier.

Next November at the HPC 2014 in New Orleans (SC14) we may get an additional data point to establish an extended period for analyzing what at the moment we consider a performance plateau trend. This additional data point will also allow us to formulate a more concrete hypothesis regarding the progression on HPC performance.

November 2014 Update

The 44th edition for the Top500 list, published November 2014, leaves no doubt regarding the performance plateau discussed in the previous pages. We have now an additional data point allowing us to established the existence of the trend documenting the stagnation in the HPC performance growth. The top 9 positions remained identical to the previous listing in June 2014.

A recent analysis[119] published by the Top500 organization just prior to the announcement of the 44th edition provides an insightful perspective to visualize the plateau trend, summarized under the following bulleted summary.

- Since 1993, 58 countries have claimed a place in the Top500 list, but only 13 of those countries have reached the top 10. From those 13, only 3 have occupied the number one position in the list, namely, United States, Japan and China.

- Of these 3 countries, only the US and Japan have a total exceeding 200 and 100 HPC systems, correspondingly. When we consider the sum of maximum Linpack performance (Rmax) reported in the Top500 list, the US have consistently exceeded the mark of 10% of the total Rmax, while Japan, China and Germany have also reached that mark as well.

- When considering the correlation between Rmax and Gross Domestic Product (GPD), this Top500 analysis plotted the US, European Union, Japan, China, and Russia. The resulting trends shows the US, EU and Japan in a similar trend, with a slight advantage on

---
119 Gary M. Johnson, Computational Science Solutions, http://www.top500.org/blog/the-big-iron-game/

the EU trend.

- However, the trend for China shows a greater commitment of GDP to Flops, with a significantly higher rate than the overall trend. Russia shows an increase in their commitment as well.

The November 2014 edition of the Top500 list[120] brought practically no change to the ranking of the top 10 HPC systems. The only new entry appears in the number 10 position, with a 3.57 PFflop/s Cray CS-Storm HPC system belonging to the US government. The US retains its position as the top country with a total of 231 HPC systems.

The number of EU HPC systems rose to 130, while the number of Asian HPC systems in general dropped from 132 to 120. The number of Chinese systems in particular also dropped from 76 to 61. Japan, however, increased its number of HPC systems from 30 to 32.

It is important to consider that the number of HPC systems is an important factor, since a multiplicity of HPC systems can definitely increase the advantage of a nation to dedicate the available multiple HPC systems to a variety of purposes, including engineering, science, and defense. A single HPC system, though it may have the performance advantage, may be utilized in a very narrow scope of tasks, and it may not contribute to a variety of purposes encompassing the areas of advanced engineering, science and defense all at once.

After determining that we are witnessing a performance

---

120 China's Tianhe-2 Supercomputer Retains Top Spot on Fourth Consecutive TOP500 List,
http://www.top500.org/blog/blog/lists/2014/11/press-release/

plateau in HPC systems, the question remaining is: what are the possible factors explaining this stagnation? I would like to offer three possible hypotheses:

First, let's consider the GPD factor and the corresponding commitment to support and sponsor the HPC industry. It is possible that we are witnessing a stressful moment in history, when the global commitment to HPC systems is globally decreasing, considering the difficult technical and logistic circumstances surrounding the HPC development beyond the current performance accomplishments. The analysis referenced at the beginning of this update section seems to support such hypothesis. Of the three top contenders, the US and Japan seem to maintain a steady pace, with only enough momentum to hold their current position in the top 10. China, on the other hand, with a discernible greater commitment on the GPD/Pflops equation, have maintained the advantage obtained since June 2013; the question is, for how much longer?

Let's not forget the dynamics leading to the event of China capturing the first place in June 2013. Six months prior to this moment, China was holding the position number 8, with the Tianhe-1A HPC system, performing at a Linpack rate of only 2,5 Pflops, behind the US, Japan and Germany. China unveiled the Tianhe-1A in November 2010, and maintained this system among the top 10 HPC systems for five consecutive cycles of the Top500 list. Since the unveiling of the new leader Tianhe-2 in June 2013, with a Linpack performance of 33, 8 PFlops, China has maintained the number one position during the last four consecutive cycles. Is China preparing to unveil Tianhe-3 on the next cycle in June 2015?

Second, let's examine the possibility of the main players attempting to channel the technical human resources and GPD support to step into the next order of magnitude in the

Cyber Reality

HPC race; the exascale. As we have already stated before in this chapter, there are compelling reasons for any nation to aspire breaking the exascale barrier, in order to obtain scientific, technological, strategic and tactical advantage. The challenges leading into breaking the exascale barrier are daunting, but since there are public statements made by several of the Top500 contenders regarding their commitment to achieve an exascale HPC system, the main contenders obviously are already exploring avenues to overcome such challenges. The one nation who successfully deploys an exascale HPC system will harvest an enormous advantage over its competitors.

After all, let's remember that there are numerous problems from the fields of medicine, engineering, and science awaiting for a solution that cannot be found with the assistance of Pflops-level HPC systems. The intricacy and complexity of these problems require a solution at the Eflops-level.

Third, is there any momentum toward the design and development of Quantum Computing? A cursory review of some reputable scientific publications[121] will answer these question with a resounding "Yes." They publish an increasing amount of research and experiments conducted toward finding solutions to the multifaceted issues associated with the design and development of a universal quantum computer. At this moment there is no universal QC operational, but an adiabatic QC optimization alternative is already operational and commercially available.[122] With this momentum in the research for QC is not infeasible to think that many Top500 contenders are currently involved in

---

121 See for instance http://www.sciencedaily.com/news/computers_math/quantum_computers/ and http://www.nature.com/subjects/physics, among others.
122 Nicola Jones, Nature, Computing: The quantum company, 19 June 2013, http://www.nature.com/news/computing-the-quantum-company-1.13212

substantial R&D efforts toward a universal QC, and part of their human resources and GPD are being redirected toward this goal.

In terms of strategic and tactical advantages, QC does offer advantages of unimaginable height to any nation capable of introducing a universal QC into their national resources. Thus, diverting resources from the HPC race into the QC race is not a superfluous strategy, but one that can open enormous return on investments.

Chapter 10. The Scope of Cyber Reality

Is cyber reality (CYRE) a universal phenomenon? Does cyber reality reach to every individual on planet Earth?

In order to answer these questions we have to enter into a dichotomy, inherited in the inquiry itself. One of the primary media required by CYRE is an information technology (IT) networked infrastructure, acting as one of the multiple conveyances used by cyber code and binary data. Since IT networks are practically ubiquitous throughout our planet, the answer to our first query is in the affirmative, and we can say that CYRE is universal in terms of infrastructure availability. However, the omnipresent condition of this IT infrastructure does not guarantee that every single human inhabitant on planet Earth capable of establishing a cognizant interface with CYRE will be willing or capable of initiating and sustaining such interfacing. Thus, CYRE is universally available, but not necessarily interfacing with every single human being at any given time in our contemporary history, either by design or by omission.

On the other hand, the majority percentage of humans interfacing with CYRE at any given time is substantially high, and we can pragmatically conclude that CYRE is indeed universal. However, we can continue our line of inquiry by raising the supplementary query: is this universal interfacing with CYRE voluntary, or mandatory?

The highest level of social organization in our modern social model requires the creation and maintenance of digital records. Therefore, every human being that is part of any community integrated with this modern social organizational model becomes an entity with a digital record, and such record is a requirement of our social structure, and thus, involuntary and mandatory in nature.

Could a member of this our modern society structure opt out of this requirement for becoming an entity stored as a digital record? Hypothetically, if this was possible, the individual opting out would become a non-entity, and as such, unable to operate within the current modern social structure. Just think what happened the last time when your digital record was absent or inaccessible in the database of the conference you were scheduled to attend? For all practical purposes, you didn't exist within the environment of that particular conference.

On one of my last trips to Europe I had three connections to my final destination. Due to a storm in the area, the airline providing my first two connections forced me to spend two days on an intermediate destination, until alternate connection flights became available. When that airline finally re-issued my new boarding passes, they omitted the third and final destination, claiming it didn't exist "because it doesn't show in our system." For the airline personnel their enterprise version of their cyber reality negated the evidence documented in my traveling schedule documentation, clearly showing my third and final destination. It took a considerable amount of time and discussion to fuse CYRE with real life.

There are primarily two issues of concern to discuss in this chapter: The first is the creation, storage and protection of personal digital records, and the second is the artificial dichotomy between personal digital records and the ontological and biological reality of the person associated with a set of personal records. In the US we maintain an official description of such digital records, known as Personally Identifiable Information (PII).[123] These digital

---

[123] See US General Services Administration, Rules and Policies -

Cyber Reality

records accompany us throughout our physical life, from the moment of our birth until the moment of our death. So, in practical terms, what is the scope of CYRE? Pretty much from cradle to grave.

The first issue arises from the fact that the accuracy of these records is linked to the degree of accuracy exhibited by the human person creating the digital record. Our society tend to forget this important factor when we nurture our unrealistic proclivity to see digital records ascribed with an inordinate degree of infallibility. We think: "if the computer says you are so and so, then you are indeed so and so." A digital record is only accurate if it accurately reflects the person associated to that record. In the event of a discrepancy, it is erroneous to assume that the person is wrong and the digital record is correct. Let us remember that the original digital record was created by a human being, and a human being is always prone to errors.

PII is data establishing a unique identifier linked to a unique person, including but not limited to a personal name or Social Security Number, along with other PII components. Instances of PII are gathered and maintained by a number of officially designated agencies, including, but not limited to, education, financial, medical, statistical, employment, etc. These same agencies are charged with the responsibility of protecting PII, but this is only a statement of intent. In real life, we all know there are numerous cases when a cyber security breach exposes our PII, usually with disastrous consequences.

What is the responsibility of the official agencies toward our digital records? And when you attempt to formulate your answer please don't limit your scope just to medical digital records. These records are indeed very important, and we

---

Protecting PII - Privacy Act,
http://www.gsa.gov/portal/content/104256

have laws protecting these medical records, but our digital records exist in many other agencies outside the medical field. On the other hand, do you think that the laws will prevent your digital records from exposure? The existing laws are in place to provide guidelines designed to protect our records. However, such laws do not guarantee our digital records are immune to exposure. Every agency holding sets or subsets of our digital records are vulnerable to intrusion, both from insiders and outsiders. The added factor in this vulnerability equation stems from the sad reality that protecting policies are not good when formulated only on records. The protective policies are only good when they are properly implemented, enforced, and updated. And the proverbial Achilles heel on these protective policies exist in the inherent weakness applied or omitted in cyber security implementations.

How technically excellent and vigilant are the hands and brains of those selected to protect the treasure of our digital records? The answer is in front of all of us: not very good, as illustrated by a never ending string of digital records exposure to unauthorized intrusions. And please remember: the percentage of exposures detailed in media reports is only a fraction of all the exposures. Why? First, because only a portion of intrusions are reported, and second, not all intrusions have been discovered yet. So, whenever you read or watch the news, you are receiving information about recently reported or discovered intrusions that actually happened several weeks or months before. By the time you learn about these intrusions, your digital records have already circulated literally around the world, or more accurately, around the underworld of crime or espionage.

According to a Government report,[124] between 2004 and

---

[124] United States Government Accountability Office, GAO Report to Congressional Requesters, GAO-08-343. INFORMATION SECURITY. Protecting Personally Identifiable Information, January

2007 there were several reported incidents of intrusions on PII, affecting different federal agencies. The total losses of PII records amount to more than 30 millions of cases, The affected agencies were found lacking compliance with all the privacy and security requirements stipulated in a government directive, and received a downgrade in their scores for E-Government progress on the President's Management Agenda Scorecard. The estimated losses regarding these identity thefts affecting US organizations reached almost US 50 billions.

So we bring accountability and penalties for the agencies found at fault, and that is part of the corrective measures. However, how were the lives of the millions of victims impacted by the intrusions into their PII? What was the personal financial impact to each one of the victims of these cases of identity theft?

The affected agencies exhibited gaps in their cyber security policies and procedures, thus reducing their ability to protect the PII entrusted to their care. The unlawful extraction of these PII can result in substantial harm, embarrassment, and at the very least, inconvenience to the victims, with enduring and traumatic consequences.

According to federal national standards,[125] the harm inflicted to individuals affected by a security breach targeting their PII is categorized in three impact levels, namely, low, moderate and high. A breach at the low level may represent mere inconveniences, while at the moderate level may include financial loss, denial of benefits, public humiliation and discrimination. At the high impact level the idamage may involve grave physical, social, or financial harm. Federal

---

2008, http://www.gao.gov/new.items/d08343.pdf
125 National Institute of Standards and Technology, Special Publication 800-122, Guide to Protecting the Confidentiality of Personally Identifiable Information (PII), April 2010.

agencies maintaining PII collections are required to report all known or suspected breaches to US-CERT within one hour.[126]

The organization publishing these standards offers a collection of guidelines on the topic of security regarding the protection of PII hosted on IT systems.[127]

The US General Services Administration (GSA) has strict rules designed to safeguard the confidentiality and integrity of PII, and such rules are well designed for the intended purpose. The problem does not reside on the available technology, or on the existing rules. The problem resides on the inevitably flawed human factor. That's right! We create all the necessary measures to protect PII, but we do not follow or enforce them, due to the human propensity to apathy, indolence, and irresponsibility.

In one case alone there were over 26 million veterans and active duty military PII stolen from a computer system at the residence of a VA employee.[128] This is not an isolated cases, but just one of the most massive PII compromise cases.

The GSA policy CIO P 2100.1E lists measures designed to protect PII. Where these measures in place in this case of compromised PII? Well, let's see. This GSA security requirement stipulates that if there is a business requirement to store PII on GSA user workstations or mobile devices such as notebooks computers, then the PII must be encrypted. Furthermore, an employee shall not physically take out PII from GSA facilities without written permission from the employee's supervisor, the data owner, and the IT

---

126 Ibid
127 The NIST SP 800 series
128 See previously cited GAO Report.

system authorizing official.[129]

Furthermore, any and all incidents involving PII data breaches must be coordinated through the Senior Agency Information Security Officer and the GSA Management Incident Response Team. So, let us ask a few questions relevant to this case.

1. What was the business requirement for this VA employee to store such massive PII collection on the portable system he took home? Was he going to review all 26 million PII over night or over the weekend?

2. Was the PII collection properly encrypted using the FIPS 140-2 certified encryption module required by the GSA CIO P 2100.1E?

3. Was this employee authorized to physically removed the PII collection from the GSA facility?

4. Did the employee obtained the required triple written authorization from his supervisor, the data owner, and the IT authorizing official?

Since the PII collection was indeed compromised, it is feasible to conclude that the VA employee was not in compliance with all these requirements. So, what happens next? Many well-intentioned and optimistic individuals will certainly join the chorus line to sing the perennial song "What we need is more training." Training regarding the obligation we have to comply with the protective policies is certainly an imperative, but is not a solution.

It is mandatory for all employees authorized to interact with PII to complete Security Awareness training and Privacy

---

[129] US General Services Administration, Rules and Policies - Protecting PII - Privacy Act, http://www.gsa.gov/portal/content/104256

Training within 30 days of employment.[130] Therefore, the particular employee responsible for this PII data breach already had the required training. We need to face the reality that training is useless when we confront the innate human tendency to act carelessly and irresponsibly, endangering the well-being of those who have entrusted us with their PII.

In cases like this we are facing the scope of cyber reality. Digital data dominates our social structure and our own lives. PII is the unavoidable format of digital records that define us as members of our contemporary society in the digital age. The organizations responsible for storing, managing and protecting our PII have designed efficient methodologies to accomplish their mission, but the human factor, the weakest link in the chain of events is always there, introducing the vulnerabilities inherent in human behavior.

The scope of CYRE is not defined by cyber systems and cyber policies; it is defined by the human element. This is the element that introduces errors in the binary code, errors in the gathering and processing of our digital data, and errors in implementing the safeguards designed to protect our digital records.

---

130 See previous GAO Report cited.

Chapter 11. The Perception of Reality

There are undeniable psychological and philosophical factors impacting our perception of the cyber reality. Our entire perceived reality is a composite of many intervening and interacting elements that allow us to process the information that we rationalize into our perceived reality. A brief examination of that cognitive process may contribute to understand the difficulty experienced by many in conceptualizing CYRE.

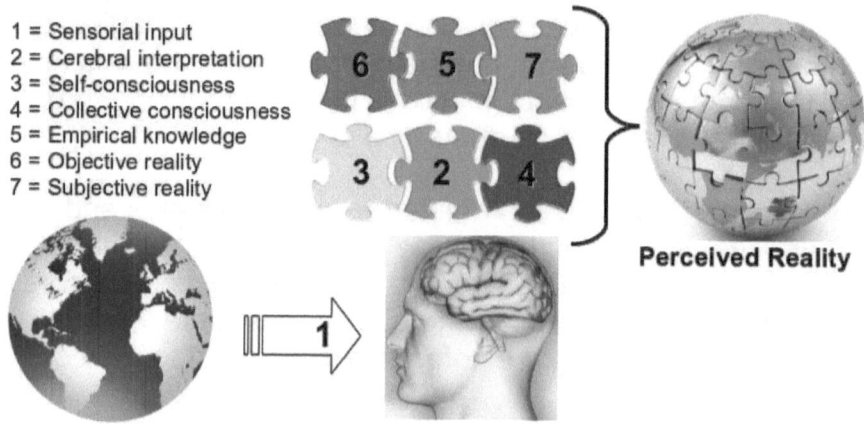

1 = Sensorial input
2 = Cerebral interpretation
3 = Self-consciousness
4 = Collective consciousness
5 = Empirical knowledge
6 = Objective reality
7 = Subjective reality

Perceived Reality

Every time we encounter a relatively novel reality we have to face the challenge of conceptualizing a concept of this reality, when faced with previously undiscovered experiences. Einstein acknowledged this challenge when he wrote[131] regarding physics and reality.

---

[131] Einstein stated that 'even the concept of the "real external world" of everyday thinking rests exclusively on sense impressions.' From "Physics and Reality" by Albert Einstein, 1936. (Translation by Jean Piccard.)
http://www.kostic.niu.edu/physics_and_reality-albert_einstein.pdf

Cyber Reality

The perception of reality is not exclusively dependent on external and sensory stimuli. Once a compendium of factors are accumulated into our subjective reality, then a thought process triggered by an empirical episode, and association process, a creative trigger, or imagination, may lead us into the articulation of a derived reality, conceptualized within our subjective consciousness, without the interaction with external stimuli.

This is exactly the case of cyber creativity, whose genesis is triggered by a thought or idea, connected with a conceptualized entity that does not yet exist. Then a cyber concept will transform this not yet existing reality into a CYRE, once the corresponding cyber code is written, developed, and disseminated, as part of the production and marketing strategy supporting the concept. This will then become a cyber entity with a tangible effect once the cyber code is installed, configured and executed within the appropriate cyber system.

The newly created cyber entity is not the result of external or sensory stimuli, but it is rather generated by the idea, imagination, and conceptual process taking place in the mind of the cyber developer. This conceptual reality generated in the mind of the cyber developer is already interacting with an existing reality, such as the cyber code being written with a particular purpose, and designed to interact with the cyber code developed for a particular cyber environment.

The tangible interaction of the new code and the previously existing code in the system becomes a mere historical incident, waiting to be witnessed at the time of code execution. However, the reality of this interaction is already created on the newly developed cyber code. Such is the case of a software code designed to patch a security gap on

Cyber Reality

an application on your computer or on your smart phone. The security weakness has already been corrected, though the tangible reality of this cyber security threat awaits for the moment in history when the user installs the newly developed security patch.

The perception of CYRE is intimately connected to our consciousness, since there is no knowledge unless self-consciousness is present. This is the crux of epistemology, the science that seeks to qualify and quantify knowledge. Epistemology[132] analyzes knowledge in relation to the methodology, validity, and scope of such knowledge. Epistemology seeks to distinguish a justified belief from a mere opinion.

When one's consciousness interacts with the external world, one must develop a conceptualization of the surrounding reality. Therefore, we advance propositions that attempt to build a concept that must be authenticated, proven, justified, until is much deeper and concrete that a simple opinion void of foundation.

The process and progression toward a viable proposition must exclude a false proposition, because it cannot be known. Why? Because knowledge requires a truthful proposition. A false proposition one doesn't believe cannot develop into a proposition one knows. Therefore, knowledge requires belief, and finally, justification. Thus, epistemology requires the presence of three interacting elements: truth, belief, and justification.[133]

Yet, it's important to highlight the fact that in this process subjectivity cannot be completely eliminated, and objectivity

---

[132] The term epistemology derives from the Greek ἐπιστήμη (epistēmē), referring to "knowledge, understanding", and λόγος (logos), referring to study.

[133] Stanford Encyclopedia of Philosophy http://plato.stanford.edu/entries/epistemology/

can only be achieved to a certain degree, without ever reaching an absolute, but only a probable certainty, contingent to the contemporaneousness of the historical milieu in which that particular consciousness exist.

The epistemic process of knowledge, therefore, resides at the intersection between belief (one's mental representation of the world), truth (the actual state of the world), and justification (the relationship between the two). The following graphic will assists us in approaching the next step in our discourse on the epistemic process.

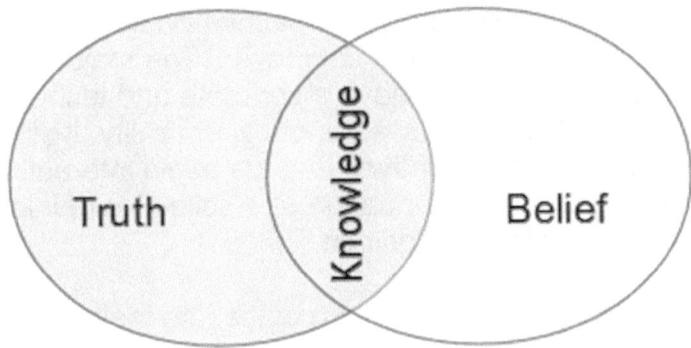

Epistemic process of knowledge

The justified knowledge then resides in the intersecting area between truth and belief. So how do we come to know CYRE? How do we come to acquire knowledge of CYRE, at a justified level where our perception of CYRE is more than just a mere personal opinion?

Let's briefly review what we recently said: When one's consciousness interacts with the external world, one must develop a conceptualization of the surrounding reality. Therefore, we advance propositions that attempt to build a concept that must be authenticated, proven, justified

I think it's safe to say that most readers remember a very

entertaining South African comedy movie released in 1980, featuring the impact that an empty Coke bottle had in the lives and beliefs of an isolated tribe residing in the Kalahari Desert.[134] The empty bottle was dropped from a plane while flying over the territory inhabited by the isolated tribe.

The perception, and the proposition that the tribe people set forward in order to understand and acquire knowledge of the unknown and strange object that fell from the sky (the Coke bottle) was structured under the influence of their collective, and isolated, cognitive background. How do we understand a new reality that all of a sudden intersects our lives? The reality of cyber intersecting our collective lives does bear a resemblance with this allegorical movie. The majority of the world population does not have a concrete and technical conceptualization of the essence of cyber reality; it simply jumps in the middle of our lives and we begin interacting with it, interpreting it on the foundation of a collective opinion, though not a justified proposition.

The tribe adopted the glass Coke bottle into their community life, and attempted to explore and elucidate its meaning and purpose. They accepted it as a gift from the gods, and it became a valuable and desirable possession, so much so that altered their lives and introduced a negative change in the social dynamics of the tribe members. The change was so devastating that the elders revised their original interpretation and declared the bottle to be an evil thing, and dictated to eradicate it from their midst.

The most interesting highlight on this allegory is the ontological dimension. There was nothing inherently good or evil in the bottle itself. The remarkable aspect in this event is the effect the glass Coke bottle introduced in the perception the tribe people had regarding themselves. Until the arrival of that object they had not realized their social principles had

---

134 The Gods Must be Crazy, http://www.imdb.com/title/tt0080801/

masked their true self. They thought they harbor no desire for personal property, but when the strange object came into their lives, all of a sudden they wanted to possess the bottle. When the desire for personal possession disrupted their social life, they decided this change was caused by the evil bottle. Actually, the bottle simply facilitated the emergence of a dormant desire.

Knowing who we really are plays a very important role in how we interpret the surrounding reality, especially with the arrival or the discovery of something new or previously unknown. Thus, epistemology and ontology are linked when it comes to our perception of reality. [135] How we come to "know" our surrounding reality has a lot to do with how we perceived ourselves, from an ontological point of view.

How do we come to acquire knowledge of CYRE, at a justified level where our perception of CYRE is more than just a mere personal opinion?

Let us then examine the road leading to the conceptualization and justification of our personal knowledge about CYRE. The propositions leading to knowledge must exist withing the environment delimited by three spheres: the personal, the procedural, and the propositional. In personal knowledge there is a combination of empirical awareness, idiosyncratic predispositions, and autobiographical facts. The procedural aspect relates to our knowledge on executing collectively known activities, such as sports and their

---

[135] There is a very interesting articled highlighting this fusion. See Gregg Henriques, Ph.D., Psychology Today, A unified approach to psychology and philosophy, December 4, 2013, http://www.psychologytoday.com/blog/theory-knowledge/201312/what-is-knowledge-brief-primer

associated rules. Propositional knowledge refers to collective knowledge on accepted views regarding the world surrounding us. [136] This latter sphere obviously plays a great role in shaping collective views on CYRE. They may be technologically and scientifically incorrect, but they are accepted as the majority view.

This process from proposition (P) to knowledge (K) is illustrated in the following diagram

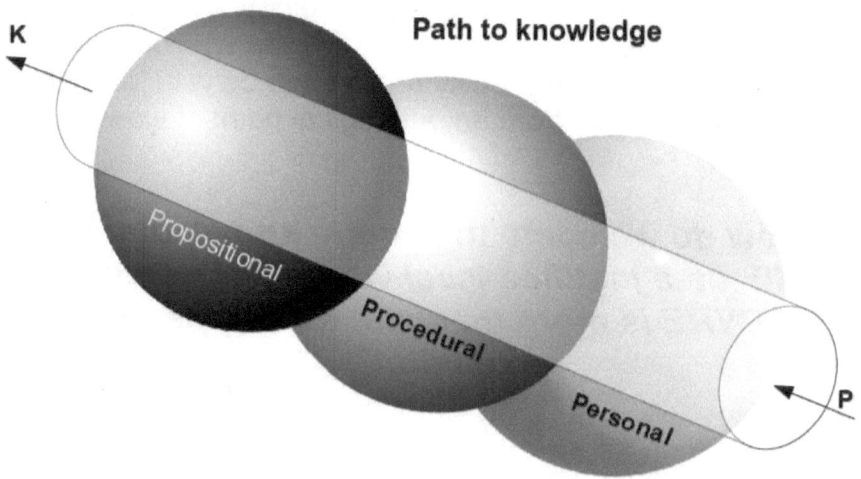

But once again we must remind ourselves that perception of our surrounding reality is at the very center of our consciousness, in the context of our sensorial capabilities. And yet, our perception transcends our sensorial capabilities as well, especially when we confront an abstract reality, or a distant one.

We have become to know that ice is "cold", along with the benefits and dangers of it. The same applies to "fire". And this knowledge has come to us through both the empirical

---

136 Ibid

and collective perception of these realities. But what can we say about the abstract nature of CYRE? Do our senses contribute to our perception of CYRE? This abstract reality is beyond the reach of sight, taste, smell, touch, and hearing. Consequently, we place a tremendous amount of trust in public opinions about CYRE, and we feed our process of "understanding" CYRE through the social context of our contemporary milieu. Thus, our perception of CYRE is in the majority of cases molded by the media, and the "knowledge" that the media may offer is at best superficial, and frequently distorted.

There is an interesting article reminding us of the centrality of our consciousness when interacting with the surrounding world.[137] The author asks: "If a tree falls in the forest, and nobody is there, does it make a sound?" Public opinion will most of the time answer in the positive. Of course it makes a sound! Really?

We tend to think that because we will perceive the sound, it is obvious that there is a sound accompanying the falling of the tree. However, the question stipulates that this occurrence happens when nobody is there. Sound is the processing and interpretations of air vibrations by a conscious entity equipped with the correspondent auditory capabilities.

When the fallen tree and its branches hit the ground there are vibrations produced by the compression of the air, with the resulting rapid air-pressure variations radiating from the epicenter of the impact at a speed of about 750 mph. Thus, when in the absence of a conscious entity equipped with a brain and ear combined system, there is no sound. All we

---

[137] Robert Lanza, Psychology Today, How life and consciousness are the keys to the universe, March 11, 2012, http://www.psychologytoday.com/blog/biocentrism/201203/the-most-astounding-thing-in-the-universe

have is the rapid radiating vibrations, but no sound is attached to them.

In summary, perception and interpretation occurs only in the presence of a conscious observer. In the absence of this observer, there is no perception and interpretation in the natural world as we conceptualize it. What then when presented with an entity that cannot be perceived and interpreted in a sensorial manner? Then we need to resort to logic and abstraction. We cannot touch, smell, hear or taste binary code. We can only see it, and we can hardly perceive it through the traditional path to knowledge briefly assessed in the preceding pages. Remember the Coke bottle in the midst of the Kalahari tribe. They processed the new information through their cognitive system, but they didn't arrive at the correct perception and conceptualization. Does our society at large provides us with a better opportunity to perceive and conceptualize CYRE? The answer is no, because society at large cannot provide the truth about CYRE. We have become global users of CYRE, but we still don't properly understand it or comprehend it.

Chapter 12. CYRE Dislocation with Real Life

There are numerous ways in which this dislocation manifests itself, leading to dramatic cases of harmful effects and tragedy. This occurs simply because of the false dichotomy cast upon the relation between CYRE and real life.

A cyber record should exist in a subservient relation to real life, designed to document and facilitate the recording of real life, but we should never allow CYRE to negate real life. Yet, the population at large has come to ascribe CYRE with a cloak of infallibility and ultimacy, and we find ourselves surrendering our human existence to the dictum that proclaims: CYRE is the ultimate reality.

This radical change in our concept of reality has led us into the illusion that the more we proclaim and adhere to our self-made CYRE, the greater our self-importance and our self-value. Thus, we erroneously conclude, the more cyber accounts we obtain, and the more we increase our exposure in the SNS, the greater our index of self-importance becomes.

This distorted sense of reality may negatively affect our personal and social set of values. A distorted sense of CYRE may lead to replacing personal accomplishments with personal cyber exposure, and to the replacement of social contributions with a distorted CYRE cacophony. This state of affairs will also replace the Cartesian dictum "cogito, ergo sum"[138] with "I post (or blog), therefore I am."

Because of the over exposure in the CYRE spectrum we

---

[138] As coined by René Descartes in 1637, in his philosophical discussion on knowledge's attainability.

have also become to equate, erroneously, that all opinions have the same intrinsic value and weight. Thus, this fallacy has also led us into the phenomenon of the democratization of opinions, as a sub-product of cyber over exposure. The principle of all men being equal does not translate into the equality of all individually-generated intellectual products.

Egalitarianism[139] cannot be translated into the realm of intellectual contributions in society. In this realm we may aspire to ensure that all members in society have access to the same platform to propose their ideas and voice their opinions. While a laudable aspirations, it is unfeasible and impractical, and a simple reality in our American government will suffice to prove this point. We live in a democracy, but do we all speak and propose in the Senate, or in the House of Representatives? While we are all considered equal in principle before the law of the land, empirically we are not all equals.

This lesson on the proper meaning of democracy has not yet mature in the minds of millions who enter the cyber spectrum to voice their minds in the cacophony of blogs and SNS fora made available for us. There is an abysmal difference at the global level between the cyber-know and the cyber know-not.[140] The population of the former constitutes only a minuscule percentage of the global population, and the sheer enormity of the latter blurs the concept of CYRE. There is a minute percentage of cyber devices' users who belong to the cyber-know category and exercise a responsible use of cyber technologies. The majority of cyber devices' users chose to simply accept the capabilities of the plethora of these available cyber devices and technologies, while showing little or no regard for the responsibility

---

139 The philosophical aspiration asserting the equality of all men regarding their access to rights and privileges in society.
140 This is a pun on the traditional categories labeled as the " haves" and the "have-nots".

associated with the employment of such cyber technologies. They tend to accept and enjoy the perceived benefits of the cyber spectrum with the same degree of disregard for understanding the technologies employed on an amusement park. Using cyber technologies is for them as simple as buying an all-pass into the amusement park, without a care for the technology they are using; they are there just to enjoy the ride.

Regarding the cyber-know-not, are they cognizant of the impact of Big Data[141] in their personal and professional life? Are they cognizant of the impact of Big Data on CYRE? Concerning the default setting of the cyber data we so carelessly disseminate, do we know the impact all this data has in the tracing of our daily life, collecting the minutiae of our interactions in the cyber global repository we casually call Big Data? Do we actually conceptualize how immensely big really is? Do we truly understand that such immense repository negates all expectation of privacy. Every thought, intention and detail we so casually disseminate in the impulsive and obsessive pseudo-Cartesian proclamation of who we are is no longer part of our private life, and we are no longer in control of it. That data now belongs to the cyber systems that recorded such disclosures, and by extension, it belongs and is under the control of the individuals who control such cyber systems.

Remember, there is no free lunch in this universe, and the so-called "free" service you subscribed in order to disseminate all the details of your personal life is actually exacting a high price from you: you have sold your privacy for the sake of exercising the privilege of telling the world who you are. Every time we interact with the cyber spectrum a plethora of metadata is collected on the system we chose to initiate an interactive session; you provide the data, and

---

141 See a cursory discussion on the nature and pervarsivess of Big Data in the next chapter.

the system collects it, and now such data is no longer under your control. The data and metadata collected by the system we chose to use now exercises control on all the information above ourselves we chose to release.

The collection and analysis of what used to be our private data is lying bare in front of the eyes of those who control access to such data, lawfully or unlawfully. When did we decided to relinquish control of our personal data? When were we seduced by the mirage of the cyber spectrum, and we so casually, and irresponsibly, decide to surrender our privacy to a cyber system? Did we think we were getting a fun and free ride?

Big data analysis can be used to examine our daily habits and preferences, and such valuable information could be used to map our daily activities with a remarkable degree of accuracy and detail. The metadata in big data is an extremely powerful tool, a very comprehensive repository of information lending a copious amount of data to the entity, legal or otherwise, with the knowledge and methodology to harvest, interpret, parse and correlate a microscopic chart of our personal lives.

What is the countermeasure to avoid the examination of our private life in the cyber spectrum? There is none. Once we submit information into a cyber system our data remains bare before the eyes of the entity owning the cyber systems storing our data. And since cyber systems do not operate in isolation, but in a networked environment, the data is disseminated among the systems pertaining to the supporting network.

A better question to ask would be: is there a mitigation strategy to reduce the degree of exposure for our personal data? Yes, since we are still in control of the amount of information we submit. We still maintain control of the elective data we choose to submit. Granted, we do not have an absolute degree of control into the submission process of data into the cyber spectrum, due primarily to the mandatory nature of the process of PII submission regulating our current society. However, we do have control into the elective portion of our interaction with the cyber spectrum. We are the masters of what we choose to submit, and slaves of what we have already submitted. Here is where a degree of wisdom and discretion plays an important role into CYRE.

What we submit into the cyber spectrum is irreversible. Once cyber data is submitted it cannot be erased, due to the process of replication of data through networked cyber systems, and the principle of ownership exercised by the entity who owns the cyber system storing the submitted data. We will remain slaves of the data we carelessly chose to submit for the duration of the cyber age.[142] Data that may appear trivial at any one time, may become a key component when parsed and correlated within the process of big data analytical process.

Our submitted personal data may appear as a field of disconnected dots to the casual observer, but to the entity capable of parsing and analyzing that data, a pattern may soon emerge, connecting the dots and offering a map of our daily life, with detailed information on our preferences and habits. The online services that offer us their cyber systems for us to enter data are the owners of that data, which is exposed to examination and study, either under the lawful or unlawful scrutiny of the eyes entitled to access and analyze such data.

None of us is in full control of our future, and whatever information we may choose to submit online today, it may become critical in the future.

---

142 For the ineluctable cessation of this cyber age, read Giannelli, The Cyber Equalizer, chapter 14

We are the masters of what we choose to submit,
and slaves of what we have already submitted.

Consequently, we can never fully control the kind and amount of data we submit to the cyber spectrum, but we can mitigate the exposure of our personal data by exercising caution and discretion today. The research and study on parsing and analyzing big data is a rapidly growing industry, and the repercussions of this progress will greatly affect both individuals and nations, with an added impact on global economy and geopolitical interactions.

Chapter 13. CYRE and Big Data

For several decades since we began storing information we arranged the stored data in homogeneous arrangements that we labeled "databases". We created databases that are designed as structured and relational databases, indicating that all components in the database are arranged in tables, and having a relational path to the other components sharing the same structured homogeneity in the same database. This relational model facilitates parsing, analysis, and correlation among the components on the same relational database. We call the orderly management of a relational database a Relational Database Management System (RDBMS). We use a special programming language, the Structured Query Language (SQL), designed to manage and communicate with the data in the RDBMS, and it's the de facto model for the storage and management of structured homogeneous data.

However, due to the explosion of ad hoc data sharing among the citizens of our networked global village, we recognized that we now have an unprecedented amount of heterogeneous and unstructured data flow that defies the traditional management systems existing in RDBMS. Thus the emergence of a new type of data, coined "Big Data", requiring a different model and management system. This need is based primarily in the unstructured and heterogeneous nature of this kind of data, generated and exchanged according to a generally voluntary and frequently trivial social model, sponsored primarily by the so-called social media phenomenon, and fueled by the now obsessive compulsion to speak about ourselves in every imaginable detail, with the deluge of information spreading liberally throughout our networked global village.

Cyber Reality

And yet, large as we may think of the copious amount of Big Data generated by the insatiable social networks, it pales in comparison with the geospatial generation of Big Data. The very complex interaction of geospatial devices and systems, including global navigation satellite systems, satellite remote sensing, sensor networks, and the ever-increasing growth of location-aware mobile devices, altogether they do generate an even larger amount of Big Data. Current records show there are 2.5 quintillion bytes of geospatial data generated on a daily basis.[143][

When it comes to speaking of Big Data there are numerous pseudo-experts who are willing to abuse this term by applying it, incorrectly, to a large amount of data. Big Data, however, is not defined primarily by the volume of the data, but rather by the composition of the data. In a nutshell, Big Data is not the same than a large amount of data. Big Data is primarily unstructured and heterogeneous data that defies the traditional model of tables in a RDBMS. A recent online article[144] illustrates the erroneous and superficial knowledge regarding the essence of Big Data, as the report speaks of intruders targeting credit cards data and characterizing this crime as Big Data theft. Credit cards data is not Big Data, because credit card information is handled by traditional RDBMS, and this information is perfectly arranged in tables of homogeneous and structured data. Another incognizant person, speaking at a recent CIO network conference, expressed his uninformed views on Big Data. During an interview this venture capitalist decided to offer his opinions on web security, a topic obviously outside the realm of his expertise. He expressed he does not considers Big Data to be all that interesting to deserve the attention currently given to it, thus revealing his shallow cognizance on the subject of

---
143 Information available from the United Nations Global Geospatial Information Management.
144 http://www.cio-today.com/story.xhtml?story_id=12100469LK34

Big Data and its importance.[145]

Big Data contains sets of information in completely dissimilar formats, from a plethora of various sources, and constantly generated at a rapid pace. This is why the proper definition of Big Data is summarized as unstructured and heterogeneous. Our information-obsessed global culture was generating, as of 2013, close to 3 exabytes of data on a daily basis. What volume are we generating as of at the closing of 2014? How many pictures, text messages, emails, blogs, trivial chats, sensors outputs, etc, are we generating today? And why do we care about gathering, storing and processing this Big Data? Because when we can correlate it and analyze it we discover that Big Data is a treasure trove for marketing, for politics, for strategies, for trend analysis, and many other uses.

Of course, this extraordinarily vast and complex data sets require an equally vast processing power; thus, enters the HPC, capable of providing the processing prowess we briefly covered previously in chapter 7. The intersection between Big Data and HPC exist at the merging of two basic needs; the intense analysis of massive amounts of unstructured information, and finding connections and patterns leading to very significant correlations. It is at this junction where the prowess of HPC systems and new analytical methodologies for extremely vast amounts of data come into place.

Big Data is not the same than a large amount of data. Big Data is primarily unstructured and heterogeneous data that defies the traditional model of tables in a RDBMS.

---

145 The Wall Street Journal, "Information Security? What Security?", Feb. 10, 2014

When we are in the presence of enormous data sets comprised of what otherwise appears as a collection of unrelated details, it is very easy to fail in identifying patterns and relations that appear indistinguishable in the midst of a seemingly chaotic array of information. This new type of analytics, dealing with Big Data, holds the promise of achieving a degree of success in meeting the challenges related to fraud detection, cyber security and insider threats. Imagine the potentials for assigning attribution to offensive cyber operations with the aid of new analytical protocols to process Big Data sets. These problems present us with the challenge to discover hidden patterns and relationships, and the discovery of dynamic solutions for tracking the discovered patterns as they manifest and develop. Just imagine the daunting task of discovering patterns and relationships hidden in the enormous amount of unstructured data generated by the overly populated circles of the social media community.

When confronted with the vast arrays of metadata generated at an enormous pace in our global village, we need to be capable of achieving pattern visualization. For this task we need the processing power that is only available from HPC systems, the working horse best suited for these demanding tasks. These exceptionally powerful systems will empower us to search and discover for hidden relationships and reveal security threats.[146]

High Performance Data Analysis (HPDA) is a tool that promise to provide us with the analysis of Big Data sets, and the unveiling of information leading to establish and implement enhanced defenses, by identifying incipient fraud

---

146 Suzanne Tracy, Editor-in-Chief, Scientific Computing, Big Data Meets HPC, 03/07/2014,
http://www.scientificcomputing.com/articles/2014/03/big-data-meets-hpc

patterns and discovering terrorists plots before they strike. We didn't connect the myriad of data bits that were available during the planning and execution of the horrendous terrorist attack we as a nation suffered on September 11, 2001. We do hope we are now better prepared to conduct Big Data analysis to curtail the new terrorist plots in the making. All the indicators are dispersed in the vast amount of unstructured data, but tools such as HPDA and others will contribute to empower us in discerning the available clues and connecting the dots to prevent future tragedies.[147]

Great benefits can be derived from the analysis of the Big Data collection proceeding from the geospatial data set, in combination with the Big Data from the corresponding sensors and the ubiquitous location-aware devices and social media streams. The analysis of this Big Data can certainly enhance the success rate of first responders in rescue and disaster scenarios. Additional Big Data from sensors in urban enclaves can also empower us to achieve improvements in traffic engineering and city developments.

Big Data is a dynamic repository of the unstructured information flowing through our complex communication networks. Now that we do have HPC systems and HPDA, this dynamic repository becomes a tremendously rich environment to find solutions and improvements to our quality of life, and to protect ourselves of the unavoidable dangers lurking in the shadows.

Our CYRE is intertwined with Big Data, since the latter can be considered a sub-product of our digital communication habits. As a sub-product we may even consider the analogy of seeing a significant percentage of Big Data as digital detritus. As such, we may even consider the possibility of seeing this sub-product as both a treasure trove and refuse,

---

[147] Steve Conway, Chirag DeKate, IDC Research, High-Performance Data Analysis: HPC Meets Big Data, March 2013

simultaneously. How, you may ask?

Let's consider this analogy from our daily lives. On a regular basis, any living individual knowingly or unknowingly sheds and disseminates biological components during any given day, including skin cells, particles, hair, fluids, etc. Likewise, the same individual may literally shed digital information about himself, including the barrage of mundane, trivial, and meaningless minutiae, disseminate liberally in the digital spectrum.

In the past we learned that even refuse and detritus can become a resource for extracting valuable components via a germane extraction process, and benefit by recycling the recovered valuable components. What is deemed worthy of recovery depends on the agenda of the recovering agent. Thus the old saying "one man's trash is another man's treasure." Are you mindful of what you put in your trash can? Once your trash is retrieve from your residential property is no longer private, and anyone can extract information about you and your habits by examining the refuse you place at the curb. The same methodology can be applied to digital detritus, and such methodology is indeed being applied today.

Presently we have the knowledge and the technology to process the deluge of digital information cast on the digital landscape, and recover, correlate, parse and analyze the information, in order to discover connections forming a particular subset of data leading to the unveiling of important clues and revelations. And this process, again, is done according to the agenda of the recovering agent. All the text messages we so liberally exchange, along with the avalanche of pictures and videos we generate about every aspect of our personal lives, collectively they all form the ever increasing repository of information that is collected in the unstructured format of the Big Data phenomenon. The expanding social contacts we maintain in the digital realm are experiencing an exponential volume augmentation via the digital replication facilitated by the social media networks, the dutiful keepers that nurture our narcissistic proclivities. Finally, the SNS have become the platform and the stage where one can tell everybody what one does and where one is at every waking moment of one's existence, so we stand tall and we shout all this information from our digital roof!

But someone may ask: if Big Data is so voluminous and complex that defies the traditional analytical tools available on a relational database, how would anyone tame this monstrous and amorphous data entity? Once upon a time we though that complex data sets such as weather and the human genome where out of our analytical reach. However, with the advent of the Chaos Theory, we are now capable of extracting valuable information from these previously inscrutable information complexes. A complex system challenges us with a vast array of interacting components and their interdependencies. But eventually we formulated a Chaos Theory that allows us to achieve a partial but predictable behavior of a complex system, by using the finite amount of information available about the system. Now, with the advent of Big Data enormous volumes of data become available, and with the assistance of HPC and HPDA we are able to increase the feasibility of higher predictable outcomes, and the unveiling of interconnected factors.

The complexity of Big Data springs from the fact that its many unstructured components can interact in so many different manners as to create a dynamic entity that transcends the sum of all the parts, adapting and evolving in response to changing circumstances.[148] A very insightful

---

[148] Geoffrey West, Scientific American, Big Data Needs a Big Theory to Go with It, Apr 16, 2013,
http://www.scientificamerican.com/article/big-data-needs-big-theory/

opinion published in a prestigious online forum discusses the parallel between the industrial age and the formulation of the laws of thermodynamic, and our information age and the imperative need for the formulation of "an overarching predictive, mathematical framework for complex systems ... [incorporating] the dynamics and organization of any complex system in a quantitative, computable framework."[149]

Big Data explosion will continue its expansion as the global digital village adopts and implements the use (and quite predictably excessive use) of the Internet of Everything phenomenon, which we will discuss in the next chapter. We need to remain cautious about the production of Big Data. Its scope and availability is not limited by geopolitical territorial definitions. Big Data is available in the cyber dimension, which transcends every aspect of life in our global digital village. Whomever has the resources and the knowledge on how to harvest this enormous field of unstructured data, has also the power to see inside the intricate details and interdependencies depicting the life and habits of the originators of such data.

---

149 Ibid

Chapter 14. The IoE Impact

The Internet community received a more mature and safe alternative for cyber communications with the arrival of IPv6, which in turn has provided the momentum for the conceptual embrace of the Internet of Everything (IoE).

Since the early 1980s we have been utilizing the IPv4, but by the beginning of the 1990s it became quite obvious that the 3.7 billion possible IPv4 addresses would not be sufficient for the rapidly growing number of potential hosts on the Internet. Accordingly, the Internet Engineering Task Force (IETF) began in 1992 a study to design a new version of the Internet Protocol.

By as early as 1994, the design and development of the new suite of protocols and standards became a reality, and in 1995 the new Internet Protocol version 6 was announced with the publication of RFC 1752, issuing the recommendation for the adoption of IPv6, as the replacement for IPv4. This recommendation was accepted by the Internet Engineering Steering Group (IESG).

In December 1998 the IPv6 base specification was published in RFC 2460, stating that IPv6 was designed as the successor to IPv4. Today, in 2014, it is quite anachronistic to refer to IPv6 as the "new" or "next generation" IP architecture, since IPv6 is already a 15 years old standard. Presently, IPv6 is widely available from industry and supported by new network equipment and operating systems.

The IPv6 Internet architecture aims to provide a mitigation to the plethora of cyber security issues plaguing IPv4, and effectively supports the rapidly increasing Internet usage and functionality. The mature design of IPv6 header format aims to keep header overhead to a minimum, and this streamlining effect supports a more efficient processing of network traffic at the level of intermediate routers, supported by a hierarchical addressing and routing infrastructure.[150]

The main security gap in IPv4 in general was the complete lack of built-in security in that pilot Internet protocol. IPv4 was never intended to be deployed as a global Internet standard, but the explosive success of IPv4 became its own demise. IPv6 is designed with security built-in, not as an after thought or a remediation technique. IPv6 uses Internet Protocol Security (IPSec), a security protocol suite integrally built into IPv6, and compliance with IPSec is mandatory when implementing IPv6. IPSec provides authentication and/or encryption for every IP packet transmitted during a network communication session. Furthermore, it provides mutual authentication between agents at the beginning of the session, and negotiation of the cryptographic keys used during that session.

Consequently, IPsec is considered an end-to-end security scheme, since is capable of protecting data flows in host-to-host sessions, between a pair of two network security gateways, or between one of these and a host. IPSec provides network traffic security along the entire network route, from source to destination.

Regretfully, the more mature cyber security design built into IPv6 is the less known advantage of this Internet technology. The aspect of IPv6 that has captured the imagination of global users is the exponentially larger number of usable IP addresses, with the added assurance that IPv6 will not suffer from IP address exhaustion, as it happened with IPv4. IPv6 was designed with a 128-bit (16-byte) base, thus providing us with a possible 340 trillion trillion trillion IPv6 addresses,

---

150 http://www.ipv6.com/articles/general/timeline-of-ipv6.htm

or 340,282,366,920,938,463,463,374,607,431,768,211,456 (3.4 x 1038). It has been calculated that "there are currently 130 million people born each year. If this number of births remains the same until the sun goes dark in 5 billion years, and all of these people live to be 72 years old, they can all have 53 times [3.7 billion Internet IP addresses] for every second of their lives."[151]

The over abundance of available IPv6 addresses has set in motion the new concept of the Internet of Everything (IoE), a network environment where each device, assigned with its own unique global IP address, becomes enable to establish peer-to-peer communications (P2P) at will. The more secure and built-in routing capabilities of IPv6 will facilitate the creation of far less complex routing tables, thus enabling a higher rate of Internet traffic performance, along with greater bandwidth availability for additional communications.

An IPv6 statistical study conducted by the IPv6 Forum extrapolates a global IPv6 Internet penetration reaching a rate of 35% by 2015 and 50% by 2020.[152] When we add this projection to the utopian concept of IoE, it is easy to see the tremendous optimism generated by the promised advantages of the IoE. However, a cautionary posture is prudent at this junction. We must ask ourselves: Is this IoE networking environment a completely benign aspiration, or does it contains the seed of self-destruction? If all available cyber devices become enabled with IPv6 and capable of immediate global connectivity on a peer-to-peer basis, how do we propose to keep the ever-present reality of any lurking malign infection outside of this idyllic social interaction?

Let's examine some of the premises of IoE. In our global quest for Internet connectivity we have by 2013 already surpassed the 10 billion mark of connected devices, and if

---

151 http://arstechnica.com/gadgets/2007/03/ipv6/
152 http://www.ipv6.com/articles/general/timeline-of-ipv6.htm

we continue the current growth trend we will very likely have 50 billion connected devices by 2020. The implications of this connectivity growth have both positive and negative components. Regretfully, we tend to focus on the former, to the detriment of the latter, simply because our emphasis is in convenience of use and multiplicity of interconnected devices, forsaking the cyber security aspect.

Is there a model to guide us into planning and organizing the complex environment of IoE? Things are being placed in motion toward a concept of interconnecting all cyber-enabled devices, and prestigious companies and influential individuals in the information and communications technologies (ICT) field are coming together to attempt the formulation and organization of the nascent IoE movement.

One such organization assisted leaders and innovators toward the goal of convening at a IoE World Forum to take place in October 2013 in Spain, to discuss all relevant aspects and formulate a working framework for the IoE quest. A steering committee has already been appointed for the IoE World Forum, with members representing businesses, governments, cities, and universities.[153] After the initial gathering the IoE World Forum convened in 2014 for the second time, hosted in the city of Chicago,[154] and a third gathering has already been announce for 2015 in Dubai, UAE.[155]

This type of initiative, seeking the utopian goal of the IoE, places us exactly in the same frame of mind we were when we conceptualized the Internet. Then we thought: now that we have computers, and we place data in them, and we use them to create solutions conceptualized out of that very same generated data, how much more wonderful would it be

---

153 http://newsroom.cisco.com/release/1152967
154 https://www.iotwf.com/iotwf2014/highlights
155 Ibid

Cyber Reality

if we can all share that data by networking all the different isolated computers?

Today we think: now we have about 10 billion cyber-enabled devices connected in the world today, and in the next 10 years we will probably grow to about 50 billion devices. If we were to interconnect all these devices to achieve a higher collective intelligence by achieving trillions of connections, how amazingly more wonderful would be connecting all these devices, working together to find very intelligent regional, continental and global solutions? Wouldn't just be wonderful? Really...?

Today we have a significant number of enthusiastic sponsors and endorsers of the concept of IoE, where many home systems such as door locks, watering devices, thermostats, ovens, lights, etc., may become "smart" devices by enabling them with sensors and Internet connectivity. Imaging the scenario where you are at work and receive a phone call from your 9-year old child that it is standing at the front door at home telling you he forgot the door key to enter the house, and he is suffering from a cold weather reaching down to 10 degrees above freezing. Even if you could jump into your car and rush back home it will take a good 40-minute drive before you can return home to let him into your house.

Now picture this alternate scenario: you connect to your cloud service, upload the picture of your about-to-freeze child, and instruct your sensor-enable cloud-connected door lock to change to unlock status upon authenticating the waiting child via facial recognition. It will take you all of 20 seconds to fix the problem, let your child into the safety and warmth of your house, and return all your attention to the project on which you were working when your child called. And in doing so you never had to take a single step away from your office.

This is the idyllic environment the IoE promises, by enabling such home devices to benefit from decision-making software residing on cloud environments, and thus allowing these home devices to be controlled remotely and adjust according to secure, intelligent and efficient designs. And there is no doubts that all these promises are perfectly possible, but...

Why but? Is there anything wrong with the concept of expanding available connectivity to any single device we utilize in our daily lives? Is there anything wrong with creating an environment where most things we use can be interconnected to enrich our lives in a more comfortable, creative, and productive way? After all, with the advent of IPv6 virtually inexhaustible pool of IP addresses, and the interconnectivity of all devices with the cloud infrastructure, why not enjoy all the conveniences offered by these innovative factors? What is wrong with aspiring for this optimized way of living?

But wait a second... Did you mention the cloud? Isn't that the new technology? No, the cloud is not a new technology; it's simply a new marketing service, designed to provide you with a network system environment where you don't have to worry about the total cost of ownership (TCO). Every individual or organization has to deal with the TCO concept, a reality of doing business, a way of quantifying the total financial impact of creating, deploying and maintaining an information technology product throughout its life cycle. Consequently, if you own an IT system, you have to calculate the initial cost, plus the costs associated with the required hardware and software protection, maintenance, and training.

The services offered by the cloud model are accessible to anyone around the world, through either public or private networks. Depending on the business model chosen, cloud

Cyber Reality

subscribers can gain access to software, applications or stored data. The cloud provider is responsible for providing all the hardware, the software, the configuration, the maintenance, the training... Doesn't that sound great?! Well, not really... What degree of control has the cloud subscriber on cyber protection? On data integrity? On data confidentiality? Would you care for an example of the cyber security issue?

An article[156] on the cloud model speaks of the Disaster Recovery issue, and explains that the solution resides on using FTP to implement a backup solution to safeguard the critical data for disaster recovery. Is there anything wrong with this advice? Most certainly. Everything is wrong when it depends on implementing critical backups using FTP, the most unsecured protocol for data transfer. There are several more secure alternatives to FTP, but if you are content with accepting the use of FTP for your disaster recovery plan, then I have a bridge to sell you...

There is absolutely nothing wrong with the IoE aspirations to a more comfortable life style enriched by the conveniences brought to our doorstep by these technologies. These aspirations are perfectly legitimate, and the feasibility of the implementation of the IoE is technically solid. Therefore, if we both want and can have the IoE, what is the real problem? The problem doesn't reside in the technical feasibility or in the legitimate aspiration to a more comfortable life style. The problem resides in human nature and human inclinations.

When we think with this utopian and idyllic frame of mind, we are today, as we were then, caught into the fallacy of forgetting the one ineluctable factor always present in human history: the malicious tendencies in our human nature, leading us to seek selfishly what will benefit us individually,

---

156 http://www.wikinvest.com/concept/Cloud_Computing

to the detriment of those around us. And in this present cyber age, those around means the entire global village.

Have we already forgotten the problem we brought upon ourselves when we decided, out of sheer convenience, to connect the critical industrial control systems (ICS) to the Internet? We designed ICS to exist as isolated systems due to their very criticality to our safety and economical well being. But one day we though: wouldn't it be great if we could interconnect these systems to the Internet? This way we wouldn't have to travel to the power plant, for instance, in order to monitor the status of all the power generating and monitoring systems. And behold, here we are now, with a cyber security problem that has become bigger than our available resources. And we did all this to ourselves, in pursue of the ever-present hedonistic drive of making everything convenient for us, even if we have to sacrifice security in the altar of expediency and convenience.

In pursuing the path of our own convenience we forgot the proclivities of human nature. We forgot what selfish pigs we can become when there is a selfish goal in front of us, and we forgot that every good intention will be match by a malicious intention if it can profit our own agenda. There is nothing wrong with aspiring to an IoE implementation, but once we have done so we will have to deal with a plethora of cyber security issues exponentially bigger than the cyber security issues generated by the ICS connectivity to the Internet.

Is this author advocating for forsaking the bright future envisioned by the capabilities of the IoE? Absolutely not. On the contrary, this author advocates for responsible aspirations, the kind where we do not rush into obtaining the perceived benefits of a new technology capable of ushering an enriched life style. We have so far proven we are incapable of defending the resources we have stored in the

Internet, now at risk in the merciless hands of malicious cyber actors. We again have shown our incompetence in defending the resources entrusted to our ICS infrastructure, and after these two disastrous decisions, now we want to bring the IoE into our lives? Yes, yes, yes, I know: this is what we want. But are we prepared to effectively defend the IoE? And if in a rush of enthusiasm the reader is ready to respond: Yes, we are prepared, how credible such statement of acquiescence actually is? What can we offer from our contemporary and collective history to prove that this time we are indeed prepared?

We are incapable of defending what we already have, and we want more? We are incapable of effectively defending our networks, both in the Government and private sectors, and we are not investing in the required number of cyber security professionals needed to reverse this pathetic status quo. Have we ever commissioned a national study to show the incredible number of irresponsible users, both in government and in private sector, contributing to the pathetic state of Internet insecurity? Obviously not, because that would be a resounding slap in our own faces. What we do instead is to repeat in a ceaseless litany the number and the kind of malicious cyber actors that are attacking our networks. This has proven to be quite therapeutic: for as long as I can concentrate in accusing the cyber attackers I don't have to face my own failures in protecting the resources under attack.

On an August 2013 blog discussing the impact of the IoE in our health, a cancer survivor in his 30s asked how user-enabled health monitoring through portable biometric devices may impact his health. We have currently available biometric devices capable of tracking blood glucose levels and sending reports to clinicians. It is expected that by 2014 we will have devices enabling users to measure temperature and heart function, hemoglobin saturation, and stress level.

Cyber Reality

Also expected are dermatological devices with tiny embedded sensors monitoring temperature, heart rate, blood pressure, blood glucose levels, and even fetal vital signs, directly from the user's skin.[157]

The initial concept of the IoE was first proposed in 1999. The IoE was conceptualized as a technology predicated on radio-frequency identification (RFID). This idea visualized all people and the objects used by them in daily life equipped with identifiers, thus allowing for computer-driven management and inventories.[158]

While acknowledging that the IoE can potentially offer undeniable beneficial factors, such benefits are substantially counteracted by the amplification of cyber security risks generated by the IoE. Consequently the IoE has been categorized as a disruptive technology in a study and its corresponding report under the endorsement of the National Intelligence Council. The IoE is given this categorization due to its capability to generate significant disturbances on critical elements of US national security.[159]

Have we ever commissioned a national study to show the incredible number of irresponsible users, both in government and in private sector, contributing to the pathetic state of Internet insecurity? Obviously not, because that would be a resounding slap in our own faces.

---

157 http://blogs.cisco.com/ioe/manage-our-own-health/#more-124074
158 Kevin Ashton, That 'Internet of Things' Thing, http://www.rfidjournal.com/articles/view?4986
159 Disruptive Civil Technologies, Conference Report CR 2008-07, April 2008

Cyber Reality

So, do you want the IoE? What is your contribution to plan a responsible deployment of this new cyber environment? What are the contributions we are willing to provide in order to deploy the required number of cyber security professionals dedicated to protect the IoE? When your personal information and sensitive personal data and medical records are bouncing from one to another cyber device interconnected in the IoE, are you sure you are going to enjoyed a concrete sense of inner peace?

Will these warnings dissuade the masses from adopting the IoE? Very unlikely, since we cannot dis-invent the things we create. Furthermore, the IoE is not really a new invention; it is simply the progression of bringing the interconnectivity facilitated by the Internet to a higher level, an exponentially expanded network of interconnected devices. Once we invent how to interconnect a few computers, and then a few millions, what's stopping us from interconnecting billions of cyber-enabled devices? In any Internet-enable nation it is hard to find a household where the Internet connectivity is limited to personal computers only. We have added our smart TVs, and our smartphones, and our tablets, and our WIFI-enabled printers, and a multitude of other similar devices.

Consequently, once we become acclimated to the concept on a myriad of our cyber-enabled devices connected to our home network, how much effort will it take to transition into the concept of interconnecting all these devices to the cloud? Doesn't the cloud sounds like the promised land where all our needs will be satisfied?

Those of us who function in this Internet era by maintaining and promoting cyber security in mind will remain very cautious about the IoE. On the other hand, the large majority that only sees the convenience and the entertainment capabilities offered by the IoE, they will flock

Cyber Reality

en masse to become absorbed into the IoE. After all, with the arrival of IPv6, we all have plenty of public IP addresses to cover all possible cyber devices we can purchase, and we are no longer limited to the private IP addresses[160] with its inherent limitations imposed by the limited availability of IPv4.

Are there concrete factors portraying the need for a cautious assessment in accepting the IoE? Indubitably yes! Let's consider for instance the extrapolation of our current global cyber security status into the imminent future of the IoE. If we have so far been unable and incapable of maintaining an acceptable level of cyber security status at both the personal and enterprise level, how can we possibly aspire at achieving such elusive goal when we exponentially multiply the number of network-enabled devices? If we are either unable or unwilling to patch our cyber systems on a disciplined and effective schedule, how do we expect to improve on our pathetic cyber security record when we surround ourselves with 20 times more cyber devices?

Do you still hold onto the myth that there are OS immune to cyber exploits, and you think that you particular OS flavor is safer that your neighbor's? This author prefers Linux-based OS, for a number of reasons already explained in this book. However, I do not hold the delusion that Linux-based OS are invulnerable. With the arrival of the IoE there will be a myriad of cyber devices that will come with Linux-based OS, simply because this OS is open source, and it is more efficient, specially on devices requiring less-resource intensive CPUs.

A reputable cyber security company recently discovered the

---

160 A public IP address is routable, thus granting the cyber device configured with such public IP address direct access to the global Internet. A private IP address, on the other hand, is non-routable, and therefore allowed to connect only to a local and private network.

presence of a vulnerability affecting Linux-based devices, and published their finding on a report[161] posted at the end of 2013. Have patches been released to eradicate the unveiled vulnerability? Have all the vendors selling the affected devices offered a solution to their customers? Have these vendors even notified their customers? Well, obviously not, and that's the reason for the publisher's article detailing the Linux-based vulnerability. According to the researchers, this newly discovered vulnerability appears to favored IoE devices as its favorite targets.

As in many other cases, the vulnerability was already patched in May 2012, but the typical indolent practice of users leaving in an alternate, make-believe cyber reality where everything is always fine, they don't pay any attention to cyber alerts disclosing newly found vulnerabilities, depending on the "well, that will never happen to me" type of attitude. If someone offers a new feature or "cool app" for portable devices, the global population is always too happy to rush into try it. If, on the other hand, a cyber alert announcing a new cyber vulnerability is published, only a handful of users will pay attention to it, and perhaps search and apply the corresponding security update.

On my home Linux-based systems, and especially the system where this book is being written, I apply an average of three cyber security updates per week. How many other users maintain a systematic schedule of security updates? How many know where to find them? How many know how to install them? So, when your inventory of cyber devices grows 400 percent, and at the present time you do not maintain a systematic patching schedule, how will you manage when this task grows 400 percent, because you

---

[161] Symantec Security Response, Linux Worm Targeting Hidden Devices, published 27 Nov 2013, updated: 23 Jan 2014, http://www.symantec.com/connect/blogs/linux-worm-targeting-hidden-devices.

have just surrounded yourself with a whole bunch of cool IoE devices? Your exposure to cyber vulnerabilities will also rise proportionally. Are you aware that most IoE devices are sold with a default user name and password? Did you change them when you purchased your device? If you didn't, those two default credentials will be exploited before too long.

One of the attractive factors of IoE devices is the low production cost. As a corollary, they become an attractive alternative to small developers who may work independently from their homes, on very low budgets. However, this autonomous developers may not have a strong incentive to integrate security in the cyber code developed for these small IoE devices. Two security researchers recently announced they discovered 19 vulnerabilities on a single IoE device, including information leakage, lack of authentication for customer data, unencrypted storage of customer data, poor password security, and poor mobile security.[162]

Concrete cases where home security and personal privacy were violated are illustrated by the 2012 incident involving 700 home security cameras. Cyber intruders gained access to the live feeds of these security cameras and made them available on the Internet, thus compromising the privacy and home security of the families involved. The manufacturing company reached a settlement with the Federal Trade Commission (FTC), resulting in the barring of the company from advertising its software as being secure. The vendor misrepresented its SecurView cameras designed for home security and baby monitoring, claiming they were "secure." Quite the contrary. The cameras operated with software lacking cyber security protection, allowing for online viewing and listening by anyone knowing the Internet address for the

---

[162] Taylor Armerding, CSO, The Internet of Things: An exploding security minefield, Apr 11, 2014, http://www.csoonline.com/article/2142722/data-protection/the-internet-of-things-an-exploding-security-minefield.html

affected cameras. Furthermore, the FTC stated that the marketer was, since April 2010, transmitting unprotected user login credentials over the Internet. This action represents FTC's first action against a marketer of an IoE device.[163]

Cases like this substantiate the position adopted by the FTC and a myriad of security experts, unanimously issuing insistent warnings about cyber security lacking in the development and production of IoE devices. Why such a big concern? Because ever since we opened the flood of IoE devices, nothing will stop us from going further and further. Enabled with the inexhaustible supply of IPv6, we will have no shortage of IP addresses to connect more and more devices, such as cars, wearable devices and mobile medical appliances, oh, and don't forget the Google Glass devices, right?

The majority of IoE devices fall into the category of embedded systems, namely, with computational capabilities that are embedded into the hardware, quite often riddled with vulnerabilities, and no intuitive or user friendly means of applying security patches to them. Why? Because their computer chips are inexpensive, and the profit margins are very low. Manufacturers combine open-source operating systems (OS) and proprietary drivers, and after the IoE device is sold there is hardly any incentive to offer upgrades or security updates until intense pressure is applied to the marketer, as in the previously outlined FTC case. Regretfully, the damage may already be irreversible.

There is an extensive number of cases where the OS on a particular IoE device may be 4 or 6 years old by the time

---

163 Federal Trade Commission, Marketer of Internet-Connected Home Security Video Cameras Settles FTC Charges It Failed to Protect Consumers' Privacy, September 4, 2013, http://www.ftc.gov/news-events/press-releases/2013/09/marketer-internet-connected-home-security-video-cameras-settles

Cyber Reality

when the device is offered on the market. An example very well documented is the case of my personal smartphone. When I purchased it the device came loaded with Android 4.1.1, even though the version 4.1.2 was already available. When I complained to the vendor and demanded to have Android version 4.1.2 loaded into my device, the vendor replied that the 4.1.2 was not yet available for users in the United States. I had to wait 4 months before the new Android version became available as an upgrade.

This example allows us to draw and follow some easy guidelines when it comes to IoE devices. Do you know what OS is installed in the IoE device you are planning to buy? Do you know it this OS is the latest available version? If not, do you know where to obtain the latest OS, and how to perform the upgrade? Another very ubiquitous IoE device is the home wireless router. One survey on these type of devices disclosed that the software loaded on them was four to five years older than the routers. And if the vendor fails to disclose this information, the users may never be aware of security updates, or may lack the expertise to implement the upgrades.[164]

The previously cited report on the discovered Linux exploit has a saga attached to it. The researchers found that if the exploit "Linux.Darlloz" finds another exploit called "Linux.Aidra" already infecting the targeted home cable or DSL modem, "Linux.Darlloz" is programmed to disable its competitor "Linux.Aidra". This should not be misconstrued as an altruistic and protective action; it is simply the action designed to eliminate the competition. The developer of "Linux.Darlloz" is utilizing the widely spread number of devices infected with "Linux.Aidra", and bring them under his

---

[164] Michael Eisen, WIRED, The Internet of Things Is Wildly Insecure — And Often Unpatchable, 01.06.14, http://www.wired.com/2014/01/theres-no-good-way-to-patch-the-internet-of-things-and-thats-a-huge-problem/

Cyber Reality

own control.[165]

Enjoying your smart TV that connects to the Internet? Obviously yes, since now you have access to streaming video and web browsing. However, did you know that a very popular manufacturer of smart TV confirmed that several of their TV models were configured to collect information on the viewing habits of the customers, and to send this data back to the manufacturer?[166] If a data breach affects the database of this manufacturer, then the cyber intruders can collect your viewing habits, and misuse this information in a variety of harmful ways. Did the manufacturer of your favorite smart TV brand informed you of this data collection? Do you know what kind of cyber protection measures, if any, has the manufacturer implemented for the protection of the collected data?

The allure of the IoE devices begins to diminish once you became aware that this new type of technology carries enormous threats to your personal and corporate privacy and security. When it comes to the government the public is always ready to jump on the wagon of "we-demand-transparency", but when it touches our personal and professional lives, we dance to a different tune, don't we? We want protection on our privacy. Just remember, the growth of the IoE will continue to increase, and the data generated by all these IoE devices will generate an unprecedented amount of data. A sober reminder; we do also have the technology to analyze enormous amounts of heterogeneous data. Just read again the previous chapter on Big Data.

In case you are not yet concerned about the IoE and your

---

165 Symantec, The Internet of Things: New Threats Emerge in a Connected World, 21 Jan 2014,
http://www.symantec.com/connect/blogs/internet-things-new-threats-emerge-connected-world
166 Ibid

privacy, let's throw into the mix the Bluetooth technology, currently engaged in evolving in order to take advantage of all the opportunities offered by the IoE and IPv6. Are you in control of your Bluetooth devices when in a crowded environment? Based on what I have seen when I'm waiting on an airport for my next plane, there are plenty of Buetooth devices wide open within the reach of my Bluetooth scanner.

The growth of the IoE is astounding. Recently Cisco released a report[167] stating that the progression on IoE devices is expected to grow from the current 14.4 billion IoE devices in 2014, to over 50 billion IoE devices in 2020! When we consider that today's global population just passed the 7 billion mark, this means we have now more devices than people, roughly about two per head. But since the distribution of these devices is not the same around the globe, that means we actually have people with more than 3 devices. I let you do the math for 2020...

There is a balance that we need to achieve when integrating IoE devices in our lives, both at the professional and personal level. We need to be willing to manage and protect those additional cyber devices responsibly. The added benefits that IoE devices may bring into our lives must remain in a delicate equilibrium with the proactive protection such devices require. Otherwise, once an exploit entrenches into the network of your IoE devices, all the IoE members of your network will fall at once. It's pretty much like having additional children; you have to acquire additional resources in order to feed them and protect them.

The bottom line is this: If a cyber device is connected to your home network, then is quite possible another person can reach into it over the Internet. The more devices you

---

167 CISCO. The network, Cisco's Technology News Site, July 29, 2013 http://newsroom.cisco.com/feature-content?type=webcontent&articleId=1208342

connect, the greater investment on your part to actively monitor and control the security settings on your devices. Are you prepared to do this? Do you know how to do it?

Chapter 15. Internet Governance

Like many other important and sensitive issues, the topic of Internet governance as been defined ad nauseam in order to support the plethora of agendas of those who seek to influence public opinion in what may appear to be perhaps the most important global decision on the survival of the Internet. This is part of the less-glamorous aspects of the truth about CYRE.

The Internet concepts, principles, and the design of its basic infrastructure were born in the minds, and through the efforts of pioneers in the United States of America. Naturally, the pristine concepts of Internet governance were also born in the US, sponsored by the very same pioneers that envisioned, developed and deployed what it has now become this flourishing global interconnected network, globally known as the Internet.

Ever since the US researchers and scientists developed and implemented the first computer network in the early 1960s, we have witnessed this network technology thriving to reach a truly global scope. Taking into account the genesis of the Internet, it is natural and obvious that the creation of protocols and policies setting the standards for Internet functionality have a US cradle as well. Consequently, the administration of the Internet infrastructure and its core resources also have an embedded US component, though the current Internet governance body has a significant international representation.

Kleinrock published the very first paper on packet switching in 1961.[168] His seminal work would eventually evolve into

---

168 Leonard Kleinrock, Information Flow in Large Communication Nets, May 31, 1961, submitted to the Massachusetts Institute of

the packet-switched network concept, where the data to be transmitted over a shared network is divided into small units (packets) and moved based on the destination address in each packet. Once all the packets arrive at the destination, they are reassembled to reconstitute the original message sent from the originating network node.

This new concept of packet-switched networking is quite different than the traditional circuit-switched technology, requiring a dedicated circuit to be established between origin and destination. The advantage of packet-switched networking is to allow multiple nodes to communicate over the same channel on a shared network.[169]

Then, when the first wide-area network (WAN) came alive, connecting two computers, one located in Massachusetts and the other in California, it became evident that remote computers could work together and facilitate the sharing of data and applications. However, this WAN connectivity demonstrated the inadequacy of the existent circuit switching system, thus pointing to the advantages and the need for implementation of the packet-switched technique proposed earlier by Kleinrock.

Licklider, another Internet pioneer, was instrumental in endorsing the computer networking project while working at DARPA, the US agency responsible for launching the ARPANET, the forerunner of the Internet. In 1969 four US computers were connected into the nascent ARPANET, with protocols developed by a cadre of network pioneers. In 1972 the ARPANET was publicly demonstrated, along with the launching of the newest application, the electronic mail,

---

Technology (MIT) as a Proposal for a Ph. D. Thesis, http://www.lk.cs.ucla.edu/data/files/Kleinrock/Information%20Flow%20in%20Large%20Communication%20Nets.pdf

169 Cisco CCNA - Packet Switched Networks, http://www.certificationkits.com/cisco-certification/Cisco-CCNA-Wide-Area-Networks-WANs/cisco-ccna-packet-switched-networks.html

the ubiquitous application known today simply as email.

The concept of the ARPANET as an open-architecture networking model was first introduced by Kahn, who also contributed with the developing of an optimized communication protocol designed specifically to support the requirements of an open-architecture networking environment. This protocol is known today as the Transmission Control Protocol/Internet Protocol (TCP/IP), the de facto communication protocol suite standard in our current Internet. The original seminal paper on TCP/IP was published in 1973, the product of a joint effort between Kahn and Vint Cerf.[170] That same year the Ethernet technology, one of the fundamental network technologies in the Internet, was also introduced.[171] The official adoption of TCP/IP as the standard among all the participating hosts in ARPANET took place in 1983, thus allowing for the segregation of the ARPANET participants into the military and non-military communities. The rapid expansion of the Internet, after the decommissioning of the ARPANET in 1990, was generated by the momentum necessary to transcend the US geographical boundaries, encompassing the entire globe.

This transition from a national to an international Internet has been made possible by the fundamental policy of maintaining free and open access to the essential Internet documentation, authored by a multitude of American network pioneers, on protocols, policies and specifications. The growing and expanding Internet also created the need for coordination mechanisms, coordination bodies, and managing boards, formed with both American and European

---

[170] V. G. Cerf and R. E. Kahn, "A protocol for packet network interconnection", IEEE Trans. Comm. Tech., vol. COM-22, V 5, pp. 627-641, May 1974, http://www.fulminiesaette.it/_uploads/foto/legame/CerfKahn1974.pdf

[171] The Internet Society, Brief History of the Internet, http://www.internetsociety.org/internet/what-internet/history-internet/brief-history-internet

representation, to assist in managing the thriving Internet activity.[172]

Twenty years after the birth of the Internet, a supplementary network service technology came into existence with the invention and deployment of The World Wide Web (WWW). The WWW service depends on the network infrastructure provided by the Internet, and the former should never be confused with the latter. The British inventor of the WWW established, and presently chairs, some of the boards and coordinating bodies regulating and managing the WWW, in order to ensure that the technologies and mechanisms sustaining the WWW continue to preserve the principle of offering its service free and open to all.[173]

One of the most important tenets on Kahn's seminal concept was the elimination of global control at the operations level. This is the very same principle that is under threat today, under the sponsorship of authoritarian foreign governments that are colluding to introduce radical change in Internet governance. Both the Internet and the WWW are threatened by the machinations of these "new kids on the block" that, arriving from the Asian continent and Third World countries into the Internet neighborhood, now want to control the ball and the game.

The Internet is indeed the undeniable gift bestowed upon the inhabitants of planet Earth by the ingenuity and industriousness of the academic, scientific and technical acumen of the US network pioneers. Does the Internet have an American footprint, ingrained in the Internet principles and goals of this 20th century American invention? Of course it does! Are we supposed to be apologetic for this American imprint into the fabric of the Internet? We invented it, we

---

172 Ibid
173 World Wide Web Foundation, History of the Web, http://www.webfoundation.org/about/vision/history-of-the-web/

designed its fundamental topology, we created its basic infrastructure, and we pioneered its fundamental protocols and policies. Are we supposed to be apologetic for this gift benefiting the millions of our planet's inhabitants?

It is precisely this American imprint on the Internet that is now questioned by the authoritarian governments threatened by the openness afforded by the global and decentralized Internet. They have discovered that the Internet at large, and the WWW in particular, have become the vehicle for the global community to learn about the agendas and practices of such authoritarian governments. They are threatened by the exposure of their activities through the open communication afforded by the free and uncensored Internet. So, in order to introduce censorship on the Internet, the only viable avenue is to disguise their agenda as a lofty and idealistic advocacy for an international governance. Translation: What they really promote is a mode of international governance excluding the influence of the United States of America, and place it in the hands of the United Nations (UN) instead.

At the present time there is a form of Internet governance with representation from the international community, under the guidance of the Internet Corporation for Assigned Names and Numbers (ICANN), directed by an internationally constituted Board of Directors. ICANN is charged with the responsibility of managing and overseeing critical technical mechanisms such as the domain name system (DNS) and Internet Protocol (IP) addressing schema. This governing body is a multistakeholder model of governance, sponsoring a collaboration process open to all Internet stakeholders. Other governing bodies include the Internet Engineering Task Force (IETF), the Internet Architecture Board (IAB), the Internet Society (ISOC), and the World Wide Web Consortium (W3C).[174]

---

174 Lennard G. Kruger, Internet Governance and the Domain Name

The list continues with the Governmental Advisory Committee (GAC), maintaining membership open to all national governments wishing participation. Presently, there are 113 nations represented, with Canada in the GAC Chair, and Sweden, Singapore, and Kenya holding Vice Chairs. Furthermore, in 2006 the UN established the Internet Governance Forum (IGF) at the World Summit on the Information Society (WSIS). The UN's International Telecommunications Union (ITU) is the specialized agency charged with overseeing communications and information technologies in general.

One of the most important tenets on Kahn's seminal concept was the elimination of global control at the operations level. This is the very same principle that is under threat today, under the sponsorship of authoritarian foreign governments that are colluding to introduce radical changes in Internet governance.

Even though the USG has no statutory authority over ICANN, it does exercise a legacy authority over ICANN, simply because the Internet evolved from a network infrastructure designed and implemented under the auspices of DARPA. With the creation of ICANN in 1998 many foreign governments began to articulate their argument opposing the USG's legacy authority over ICANN and the DNS. In 2005 the UN's Working Group on Internet Governance (WGIG) released a report stating that ICANN would either be replaced or made accountable to a higher intergovernmental body. US officials counteracted by declaring that the USG will not relinquish the control and administration of the DNS

System: Issues for Congress, November 13, 2013

to any international body.[175]

Ever since then the debate is kept alive by many foreign nations, specially by the relentless efforts of the BRIC coalition formed by Brazil, Russia, India and China. This coalition is preparing to foster their adversarial position on Internet governance during the upcoming Global Multistakeholder Meeting on the Future of Internet Governance (NETMundial), scheduled to be hosted by Brazil in April 2014.[176] The agenda of NETMundial will focus the efforts of the BRIC coalition on forging their own principles of Internet governance and the proposal of a guide for evolution of Internet governance. As the author of this book, am I allowed to express a sarcastic comment? Of course I am; that's why I'm the author (unless the BRIC has a different opinion?). I am certain the BRIC will have a resounding success in carving their own version of Internet governance guide, since their contribution to the design, creation, and implementation of the Internet is so well documented in the history of the Internet we just reviewed in the preceding pages, right?[177]

Among the many brave and enlightened voices sounding the alarm against the machinations of the collusion between ITU and BRIC, there are two in particular that provide a concise and candid exposé of their stratagem. One of them explains that freedom of speech and freedom from censorship will be lost under any form of Internet governance exerted by the ITU. The BRIC countries and their sympathizers are leveraging UN approval (through the ITU) in order to wield government control over the Internet, seeking to subjugate

---

[175] Ibid
[176] Internet Corporation for Assigned Names and Numbers (ICANN), 11 January 2014, http://www.icann.org/en/news/announcements/announcement-11jan14-en.htm
[177] The author wrote this comment during March 2014, prior to NETMundial.

the threat to the stability of their authoritarian governments, placed at risk by the anti-censorship principle in action through the Internet.[178]

The other voice speaks of the ITU as that 150 year-old bureaucracy originally charged with establishing telegraph and telephone standards, but presently lacking any role in Internet management. ITU relentless attempts to infringe into Internet governance have met with global condemnation, including major industry representatives and engineering organizations, such as the international Internet Society. The dangerous coalition of repressive national governments, led by the BRIC coalition and the ITU, represent a real and present danger to the survival of an open Internet. The plan of the BRIC-ITU collusion is to use the UN as the cover to implement and enforce repressive censorship by nation-state members.[179]

Let's examine a few cases of Internet censorship implemented by the very same nations that are planning to eliminate ICANN and the USG from the body administering Internet governance. The Chinese Communist Party has systematically implemented means within their borders to block web sites they deem undesirable for their national policies, by implementing technologies that limit freedom of expression. Furthermore, the Chinese government maintains a tight control in monitoring the online behavior of their citizens and enforce the government rules, and dealing ruthlessly with those who break them. In 2009, the Chinese writer Liu Xiaobo received a sentence for 11 years in prison

---

[178] Susan Crawford, Harvard Kennedy School, Internet Governance: Threats and Opportunities, December 2011, http://www.technologyandpolicy.org/2011/12/15/internet-governance-threats-and-opportunities/#.Ux_OT9sy3CI

[179] Larry Downes, Forbes, Why is the UN Trying to Take over the Internet?, August 2012, http://www.forbes.com/sites/larrydownes/2012/08/09/why-the-un-is-trying-to-take-over-the-internet/3/

for co-authoring an online manifesto opposing the Chinese authoritarian government.[180]

The Freedom House 2013 report highlighted China, Iran, and Saudi Arabia as the nation-states implementing very comprehensive filtering and blocking technologies, accomplishing an effective access denial to thousands of web sites. The report also accentuates the increase in the practice of using paid pro-government commentators, covertly hired by government officials to manipulate online discussions by trying to smear the reputation of government dissidents. Russia, China, and Bahrain have engaged in this stratagem for a number of years, and recently joined by Malaysia, Belarus, and Ecuador.[181]

In many authoritarian governments there is an important condition favoring ICT censorship, due to the fact that in such countries the government has ownership of the main ICT firms, Consequently, the state is exempt from the burden of producing legal warrants prior to implement surveillance and censorship against their dissidents. Furthermore, these states can expeditiously launch disruption and blocking actions against Internet sites and mobile communication services in order to preserve government interests challenged by dissident views.[182]

In 2011 Egyptian authorities plunged the country into Internet isolation in an effort to curtail internal political unrest. The government of Syria has launched not one but a series of Internet isolation episodes between 2011 and the present. When considering the Internet connectivity schema of Syria, with a single government-controlled gateway into the global

---

[180] The Economist, China's Internet . A giant cage, Apr 6, 2013, http://www.economist.com/news/special-report/21574628-internet-was-expected-help-democratise-china-instead-it-has-enabled

[181] Freedom House, Freedom on the Net 2013. A global Assessment of Internet and Digital media, October 3, 2013

[182] Ibid

Cyber Reality

Internet, and the documented systematic manner in which the border gateways were withdrawn, there is only one logical conclusion; the power controlling these border gateways is the power implementing the Internet isolation.

In Venezuela, the main national ISP severed connectivity to the global Internet during the presidential election in 2012, and India and China disabled text messaging on mobile communications serving particular areas affected by civil unrest. Russia has became the enabler for the production and dissemination of ICT surveillance technologies (SORM), adopted by neighboring countries such as Belarus, Uzbekistan, Kyrgyzstan, Kazakhstan, and Ukraine, among others.[183]

The Chinese government has implemented the world's most comprehensive and invasive Internet filtering system, making it impossible for anyone in China to obtain information on politically sensitive subjects such as the Tiananmen Square protests, or human rights and political reform. The filtering system also protect their political leadership from exposure of any embarrassing information about them.[184]

On the same year, following riots in Xinjiang, government authorities isolated the region from accessing the Internet for about a year. The tools created to monitor online behavior are transcending the Chinese borders, and being offered to other authoritarian nation states. Huawei and ZTE, two important ICT companies, are the leading suppliers of internet and telecommunication hardware and software. Their client list includes a number of states in Central and South-East Asia, eastern Europe and Africa. All these

---
183 Ibid
184 Eric Schmidt and Jared Cohen, Web censorship: the net is closing in, The Guardian, Tuesday 23 April 2013, http://www.theguardian.com/technology/2013/apr/23/web-censorship-net-closing-in

Cyber Reality

nations share the authoritarian model of imposing constricting political and technological control over their populations. China maintains an alliance with these countries, and especially with Russia, in their plan of disputing the current Internet governance model.[185]

A recent 2014 news article[186] lists the top 10 nation-states enforcing Internet censorship:

1. North Korea, where all websites are under government control.
2. Burma, filtering e-mails and blocking web sites exposing human rights violations or dissident views.
3. Cuba, with government controlled "access points" and heavy monitoring activity.
4. Saudi Arabia, with over 400,000 blocked web sites hosting views contrary to the beliefs of the monarchy.
5. Iran, where bloggers must register at the Ministry of Art and Culture, and dissidents suffer jail terms.
6. China, implementing the most rigid censorship program.
7. Syria, where dissidents are arrested, and cyber cafes must enforce strict reporting rules to authorities.
8. Tunisia, where all ISPs must follow rigid reporting rules to the government.
9. Vietnam, implementing invasive monitoring rules and blocking of web sites criticizing the government.
10. Turkmenistan, where the only ISP is the government, monitoring all users' activity.

A truly historical announcement was published by ICANN on 14 March 2014, while the author was writing this chapter. ICANN's statement was released immediately following the historical announcement by the USG, stating its readiness to

---

185 Ibid
186 USA TODAY, February 5, 2014, Top 10 Internet-censored countries, http://www.usatoday.com/story/news/world/2014/02/05/top-ten-internet-censors/5222385/

transfer its stewardship on the important Internet technical functions to the global Internet community. This transitional process includes the USG current responsibilities of its procedural role in administering changes to the DNS authoritative root zone system, the database with the names and addresses of all top-level domains, the unique identifiers registries for domain names, IP addresses, and protocol parameters.

The maturity of ICANN as an effective multistakeholder organization is recognized by the USG, and ICANN received the request to convene the global community with the task of developing the required process to transition stewardship from the USG to a properly organized global community consensus-driven body, including private sector, civil society, governments, and other Internet organizations.[187]

This transitional process should not be misconstrued as a cessation of the responsibilities of ICANN. The transition process only includes the stewardship over ICANN responsibilities, but ICANN itself remains as the organization tasked with the role as administrator of the Internet's unique identifier system. These functions remain a critical component in the overall task of maintaining a global functioning Internet entity. ICANN will continue performing the vital technical functions required to sustain the availability and security of the Internet infrastructure.[188]

The transitional plan should be in place and ready for activation by the time the current ICANN contract with the USG reaches its expiration date in September 2015. At that time it is expected to have a clearly defined process to adopt a global multistakeholder stewardship of ICANN 's critical

---

[187] ICANN, Administrator of Domain Name System Launches Global Multistakeholder Accountability Process, Press Briefing Scheduled with Board Chair and CEO, 14 March 2014,
https://www.icann.org/en/news/press/releases/release-14mar14-en

[188] Ibid

responsibilities. Toward this goal a community-wide exchange will initiate the transitional process during ICANN's 49th Public Meeting, scheduled for March 23-27 in Singapore, where all global stakeholders are welcome to participate.[189] This is expected to become a truly momentous event, with historical significance.

Some reputable press articles have reported the ICANN announcement accurately, even though the same cannot be said of all press reports, especially those who prefer sensationalism over the truth, and distort the meaning of the announcement, either because of ignorance on ICANN's functions, or in order to boost their profitability in disseminating the related news.

The transition process raises concerns among global businesses depending on the availability and reliability of a functioning Internet. The USG has resolutely stated that no government-led organization will be allowed to subjugate ICANN, and consequently, the ITU will not be allowed to assume control of Internet governance.[190] The multistakeholder model must be preserved and enhanced by the transition process,[191] in order to maintain the original tenet of an open and uncensored Internet.

So, in the aftermath of NetMundial, what can it be said? Where there any earth-shattering developments or unexpected surprises? According to the summary coverage of a major European media, none of the aspects of the NetMundial outcome are unexpected, considering the agenda pervading NetMundial goals, already analyzed in

---

189 Ibid
190 Edward Wyatt, U.S. to Cede Its Oversight of Addresses on Internet, March 14, 2014, http://www.nytimes.com/2014/03/15/technology/us-to-give-up-role-in-internet-domain-names.html?_r=0
191 Joseph Menn, U.S. government aims to shed control of Internet addresses, March 14, 2014, http://www.reuters.com/article/2014/03/15/us-usa-internet-domainnames-idUSBREA2D1YH20140315

this chapter. On a recent article published by this media[192] they state that Russia and China, along with two other supporting nations, are advocating for the UN to take leadership in the subsequent deliberations regarding Internet governance. This outcome is not a surprise, since is the focus of Russia and China's agenda. This also shows the contemptuous disregard these two countries have for the terms announce by the USG, stating that no government-led organization will be allowed to subjugate ICANN, or allowed to assume control of Internet governance.[193] The resolute insistence of Russia and China highlights the unwavering agenda of these two nations seeking to exercise control over the Internet by using the UN as their proxy.

On April 14 NetMundial published a draft outcome document,[194] with recommendations arranged around two main sections: Internet Government Principles, and Roadmap for the future evolution of the Internet Governance Ecosystem. On April 21 the USG responded with a series of comments to the second draft, indicating that this NetMuldial document requires some critical changes. In general, there are two primary areas requiring amendment. The language addressing human rights in the current reading is blurring the line of separation between long-established rights, and other values addressing a collective vision of the Internet. This blurring effect represents a risk of undermining these long-established rights. The other main area concerns the focal point of NetMundial. The draft should focus on encouraging global cooperation in dealing with the expanding cyber crime and cyber security issues instead of promoting controversial additional global treaties and agreements.[195]

---

192 Leo Kelion, Future of the internet debated at NetMundial in Brazil, BBC, 23 April 2014, http://www.bbc.com/news/technology-27108869
193 See previous footnote 146
194 http://document.netmundial.br/
195 USG Comments on Second Draft NETmundial Outcome Document, April 21, 2014,
http://www.state.gov/documents/organization/225225.pdf

Collectively, the principles and the roadmap outline should serve as a guide for the future discussions to come on the road to a mature transitional plan, open for discussion among the existing bodies. This draft should not encourage the creation of new requirements or mechanisms altogether. Some of the principles in the NetMundial draft contains language that is not present in human rights instruments endowed with international acceptance. The presence of the language in question gives the impression that NETmundial is endorsing the creation of new rights.

The NetMundial draft also attempts to introduce language that is not present in the International Covenant on Civil and Political Rights (ICCPR), from the Office of the High Commissioner for Human Rights. Language in a document of this magnitude should provide a guideline for the NetMundial draft, and yet, the draft chooses to create a new lexicon. Why? If we are speaking of human rights, and there is a high level document addressing human rights, such as the ICCPR, why not using its language? This is a very interesting situation. Let's recapitulate what NetMundial allegedly was set to do.

I previously stated that the BRIC coalition was preparing the stage to foster their adversarial position on Internet governance during NetMundial. Their agenda was to focus their efforts on forging their own principles of Internet governance, and to advance the proposal of a guide for evolution of Internet governance. The BRIC countries and their sympathizers are leveraging UN approval (through the ITU) in order to wield government control over the Internet. And this is exactly what they tried to do.

It is very strange that BRIC claims they want to use UN as a guiding body for Internet governance, but if they cannot find the appropriate language for their purposes in the official UN document addressing human rights, then they will change to whatever terminology serves their hidden agenda. They are expecting that their demagogy will serve as a smoke screen so the attendees will not notice the language change in the midst of their empty rhetoric. It's the oldest trick in the world; tell the crowd what they want to hear, while hiding the real goals in stealthy language disseminated through official documentation. Everybody heard what it was said, but did everybody read carefully what was written in the official draft?

The USG has resolutely stated that no government-led organization will be allowed to subjugate ICANN, and consequently, the ITU will not be allowed to assume control of Internet governance. The multistakeholder model must be preserved and enhanced by the transition process, in order to maintain the original tenet of an open and uncensored Internet.

So, allow me to address every reader by underlying the main issue in this chapter. Whatever iteration of CYRE you may have articulated in your personal and professional life, there is a dark side that we must all face. The objective reality of the purpose and goal of the Internet, as envisioned and implemented by the early American pioneers, and safeguarded during the last 25 years, is at risk of disappearing. Authoritarian nation states, under the cover of the UN and the ITU, are persistently launching frontal attacks on the current Internet governance model. They have invested the last 10 years in reinventing their adversarial machinations under different names, but their final objective remains the same: to eliminate the historical role of the US sponsorship and protection of the pristine tenets of Internet openness and freedom of speech. These tenets are contrary to the agendas of these authoritarian opponents, but they disguise their true motives by appealing to international support under the false pretense of seeking global representation and security within the Internet environment.

The current Internet governance model already has global representation and promotes Internet security, so why do these repressive nation states continue their masquerade? Because they know that they may eventually dissuade the international community of accepting the fallacy that eliminating the US role from Internet governance is better for everybody.

This is the dark side of CYRE: one day you may wake up to a different reality, where all the freedom, availability and openness you have enjoy all these years, under the current Internet governance model, have vanished. Your personal CYRE, predicated on the availability of these very precious elements, will dramatically changed, if we allow ourselves to be seduced by the repetitive and deceiving mantra sponsored by the UN, the ITU, and the BRIC members and their sympathizers.

Chapter 16. CYRE & Global Economy

The CYRE we face on an everyday basis has been forged in the last 50 years by an economical phenomenon formulated during the 1970s; the globalization of manufacturing. This recent development in the economic schema consists of integrating manufacturing processes from international sources, and even though this global manufacturing model was thought as an advantageous strategy in its beginning, it has introduced a dichotomy factor into the CYRE. For a while we rejoiced in achieving the goal of minimizing manufacturing costs by selecting the most advantageous workforces opportunities. Now, however, we have come to the realization that we have exposed ourselves to become vulnerable to all the cyber vulnerabilities introduced by international manufacturing sources who, under a national agenda, may pursue to exploit their clientèle.

Ever since the world greeted the Industrial Revolution age, many business models were developed by the competing companies in order to maximize their market share and profits. Beginning during the 1970s and forward, a new business strategy began to take shape. One such strategy started focusing on discriminating between core business components and ancillary ones, while seeking a profitable way to outsource the latter ones in order to introduce cost saving advantages into the business products and services.[196] This strategy sought to increase business profitability and flexibility, and when it began approaching a degree of maturity as a business model during the 1990s, it became the precursor to the current globalization model we practice today. As in many other strategies, this new model

---

[196] See additional details on this trend on Robert Handfield, A Brief History of Outsourcing, June, 2006, http://scm.ncsu.edu/scm-articles/article/a-brief-history-of-outsourcing.

created advantages and vulnerabilities, and today we are confronted with the imperative to maximize the former, while mitigating the latter.

The evolution of this nascent business model gave birth to the axiom that developing strategic partnerships was more important than holding the ownership of core business components. Consequently, outsourcing became the strategy of contracting out major functions to external service providers with the capability of offering specialization and efficiency on specific functions, while entering into a mutually profitable business partnership. This outsourcing model was considered the answer to the need for reduction in operating costs, access to specialized capabilities, and sharing of business risks with business partners, among others. However, in adopting this business model that developed into the globalized economy we have today, we erred in anticipating the loss of control on the production and distribution of the products and services we were so willingly outsourcing. The cycle of multilevel outsourcing strata became so complex that today we can hardly determine the integrity, quality, and reliability of the products exchanged through the global supply chain. Today CYRE is interwoven into a complex gossamer of almost untraceable layers of partnership, and into an intricate web of industrial, scientific, and defense espionage.

When we became enthralled with the possibilities of increasing business profits by establishing partnership agreements and outsourcing the manufacturing and generation of our products, we became oblivious to the fact that we were also relinquishing control on the integrity and reliability of our products and services. This issue is at the very crux of our current cyber risks and vulnerabilities. Do you know who manufactures your cyber devices? Do you know how many different components are part of the assembly of your cyber devices? Do you know what are the particular agendas sponsored by the myriad of outsources interwoven in the final assembly of the cyber devices you have in your network, at the enterprise or personal level?

Do you know who manufactures the laptops we use, that you assumed are American products because they are marketed under a brand name that sounds American? Do you know who is writing the source code of the anti-virus product you have installed in your cyber devices, at work or at home? Do you know who manufactures the embedded cyber systems you have deployed at home or at work, and who wrote the embedded cyber code on them? Do you know who manufactured or assembled the routers, switches and network peripheral devices you deployed in your enterprise network? Do you know if your enterprise procurement office has the cyber professional expertise to discern and discriminate between the available sources for these devices? Do you know who is controlling the data flowing through the foreign ISP network you are contracting to provide connectivity services for your overseas personnel? Do you know what degree of direct and discrete control this foreign ISP has over your data traversing the ISP network?

Have you ever ask yourself these questions? Perhaps when you do, or when you begin to do so, you may start gaining awareness of the complex environment of our present CYRE. It is neither a simple nor a comforting reality; CYRE is a nonlinear reality,[197] and since we embraced the global economy model with its outsourcing spawn, now we must confront the spectral figure of the supply chain threat (SCT) on a constant basis. SCT is a persistent threat that requires careful and strategic attention and monitoring. Is this warning just the expression of an overly paranoid concern sponsored by this author? Definitely not, since our government has recognized the importance of this threat, and dedicated ad hoc resources to monitor and counteract it,

---

[197] Giannelli, pp. 69-70, 109

by creating the Office of the National Counterintelligence Executive (ONCIX).[198]

The ONCIX focuses primarily on four of the main components on the threat at hand, including SCT, cyber security, cyber espionage, and the internal threat. SCT is the threat component set in place by the current globalist economic model that has facilitated the placement of critical links in the supply chain directly under the control of US adversaries. The SCT affects both the commercial and the defense environments, while exploiting the interdependency between these two.

Since we relinquished control over the supply chain we discovered that achieving reliable attribution has become an extremely difficult task, and unearthing the true identity of the adversary poisoning the supply chain is an exceedingly rare accomplishment. Our adversaries use their access to the supply chain in order to acquire sensitive data on innovative technologies and admission into sensitive networks and systems, along with the insertion of counterfeit cyber devices, designed to maintain permanent remote access, and when necessary, inject performance degradation into US systems.

As a nation, our dependencies on cyber technologies present an irresistible target for our adversaries, since through exploitation of cyber systems they can collect intelligence on our national strategies and operations, and they can also introduce disruption and degradation effects on the cyber systems and technologies on which we depend. Our adversaries, including foreign intelligence entities, a variety of adversarial cyber actors, terrorists and cyber criminals, can rely on cyber operations against the US because of the low cost of cyber operations, paired with a low risk and high return of investment. The preparation for

---

198 http://www.ncix.gov/

the attack environment can be achieved very expeditiously, and there is a myriad of technical tools and available network infrastructures to mitigate the risk of attribution.[199]

Today we have to contend with the increasing intrusive activity from foreign intelligence services, private sector espionage, and cyber criminals that can operate in collusion against the national interests of the US. These adversaries focus their nefarious efforts in compromising technological developments and research data considered critical to our national security.

The threat of espionage is not limited to traditional adversaries, such as China and Russia, but to allies as well as insiders. According to US government records, the cases of Chinese espionage prosecuted by the FBI reached an all high peak in 2010. This, in addition to several cases of Russian elements infiltrated in our country, with the mission of collecting information on important political, military and economical matters.[200]

Counteractively, we as a nation can wield the tools afforded to us through counterintelligence procedures, in order to mitigate the combined threat presented by foreign adversaries, allies, and insiders. This last group is empowered to become the agent capable of inflicting the most damaging harm to our national security, along with diminishing our scientific, technical, and military superiority. Insiders already have been granted access to restricted and sensitive information, and considering the enabling capabilities offered by digital data duplicators, insiders can actually extract more data than in any other previous period in our national history.

---

199 Additional details on this issue are available directly from the public ONCIX site, at http://www.ncix.gov/issues/supplychain/index.php
200 Ibid

Cyber Reality

At this point of our discussion on cyber collection and espionage as performed by our national adversaries, volumes have been written about the large amount of information that is methodically and persistently extracted from research institutions, Cleared Defense Contractors (CDC), industry and military institutions. But in addition to engaging in discussing an undeniable fact, are we simultaneously increasing our demagogic appeal in order to obfuscate our lack of prevention in protecting our sensitive data? Simply as a rhetorical question, let us ask ourselves: Do we only have the right to create data but not to protected it? If I store treasures in my residential place, do I leave doors and windows unlocked? Are we simply being naïve, or rather guilty of the utmost irresponsibility?

What do we do with the data created by R&D, generated at a substantial cost? Do I store it in protected cyber storage devices, or do I leave it in nominally protected cyber storage devices? When I produce costly research data, do I make it publicly available, only to lament when our adversaries collect it from those publicly available places? Do we really have a viable expectation of maintaining that data as private while making it available on public places? If we do maintain such expectation, we have then a very twisted and distorted sense of CYRE. Digital data, when is meant to remain our private property, should never be place on a publicly accessible place. When the flowers offer their openness in Spring, bees do come. When their openness is no longer offered, bees do not come. This is an extremely simple principle, right? So, why do we hold onto our irrational expectation of privacy while exposing our precious data in open networks and cyber storage devices? Are we being simply disingenuous, or possessing less intellectual acuity than vegetables?

It is indeed refreshing and encouraging that we are making great progress in enacting legislation that allows us to prosecute cyber criminals responsible for the theft of proprietary or sensitive cyber data. Some of the most recent efforts are highlighted in a newly published governmental strategy,[201] but these legislative improvements allow us only to establish the legal responsibility and penalties sanctioned by law upon the culprits. However, such legislative measures do not prevent the theft of cyber data. Our cyber protection cannot come exclusively from wrapping ourselves in legislative documentation. We need to empower ourselves with the counter cyber tools and counter cyber assets who hold the empirical and technical knowledge to prevent the theft of cyber data in the first place. We do have such human assets, but are we using them? Do we know how to enlist their professional services? Are we actually enlisting them to protect us from the theft of our trade secrets? The answers to these few questions are being shouted from the daily news. Only a minute percentage of the cyber thefts are being disclosed by the large amount of victimized entities affected by cyber theft.

We can certainly rejoice that our newly enacted cyber laws will provide the legal foundation to punish cyber criminals, but it is exponentially simpler and more economical to stop them from accomplishing success in their cyber predation. A deluge of hyperbolic but empty statements will not suffice to protect us from the cyber predators surrounding us. This issue in front of us is one of the principal components of CYRE. Our cyber data, economically and strategically very valuable, is being siphoned away in front of our very own eyes, and the best we can do is to focus on creating the legal foundation to prosecute the thieves? How about finding the way to stop them from stealing our costly cyber data? Isn't the right to self-defense one of the most fundamental and inalienable principles in any society? Do I have to wait for the enactment of a law before I can defend

---

201 Administrative Strategy on Mitigating the Theft of U.S. Trade Secrets, February 2013

myself, before I can defend my property? Once again, I must reiterate the question for those who may dare to answer in the affirmative: what kind of twisted and distorted CYRE do we hold as individuals, and as a nation?

What is one of the other favorite demagogic and hyperbolic statements we are broadcasting in seeking the proposed goal of mitigating the stealing of our cyber data? The one that says: let us educate the population regarding the threat of data theft. And while conducting this 101 class on mitigating data theft, what are we doing to stop the hemorrhage? Well, do we think that by immersing ourselves in becoming aware of the data theft we are actually mitigating the problem? Are we that gullible and naïve? Sure, let's sit down in the classroom and learned about the data theft while we bleed. Do we really need expensive reports to Congress stating in a highlighted bullet that two adversaries from the Eurasia continent will continue conducting cyber adversarial operations against US targets? Are we in the business of stating what is painfully obvious? And yet such type of reports are being written.

How many of those who commissioned the report, and those who wrote the report, and those who read the report were actually involved in mitigating the bleeding? Have the demagogues consider using the services of those of us who already know about this data theft problem, and we know how to mitigate it? In times of crisis there is no place for complacency and relaxation. Let's conduct cyber data theft 101 and promote threat awareness after we have mitigated the crisis, not while we are in the middle of the crisis. There are times to talk, and there are times to act. This is the time to act. Let's talk later.[202]

---

202 For a summary of cases the reader may consult the report "Administration Strategy on Mitigating the Theft of US Trade Secrets", February 2013, Annex B: Summary of Department of Justice Trade Secret Theft Cases, January 2009 through January 2013.

An ONCIX report[203] on this topic ends with a series of guidelines arranged in six sections containing 19 bullets on how to protect sensitive data. They contain sound advice, but they all concentrate on procedures and the use of software. What is lacking, and this is symptomatic, is any advice about incorporating a small team of cyber security experts (one or two members will suffice), to conduct prevention and detection of data theft. The last bullet suggest to contact ONCIX or the FBI for assistance on strategies for data protection. This is also symptomatic as well; these two organizations do not own the exclusivity of expertise on cyber protection, which is purely technical and available from many reputable sources. The best suited strategy to defend against data theft is an internal small team of cyber security experts, certified and qualified, operating from inside the enterprise network, equipped with vast knowledge of the internal network topology, and with empirical prowess forged in many actual cyber battles with the adversary. These are the true cyber warriors, and they are never born overnight.

Our cyber protection cannot come exclusively from wrapping ourselves in legislative documentation. We need to empower ourselves with the counter cyber tools and counter cyber assets who hold the empirical and technical knowledge to prevent the theft of cyber data in the first place. We do have such human assets, but are we using them?

---

203 ONCIX, Foreign Spies Stealing US Economic Secrets in Cyberspace: Report to Congress on Foreign Economic Collection and Industrial Espionage, 2009-2011, October 2011.

Cyber Reality

Why did I say that this gap in defense strategy is symptomatic? Because an overwhelming majority of germane reports are written by individuals whose knowledge of the technical cyber intricacies is superficial at best, and lacking the empirical acuity of cyber defense, they concentrate on procedures, excluding the paramount factor of human cyber expertise. Procedures are definitely necessary, but only as a complementary factor, not as the driving force formed by human cyber security experts, the true determining factor required to implement an effective cyber defense strategy.

No outsider, regardless of their good intentions, can outperform an internal team in preventing and combating a cyber attack, simply because an outsider team doesn't know the network topology as the internal cyber security team does. This internal team is the only means of interacting with a cyber data theft attack in an intelligent, analytical, proactive, and strategically and tactically sound counteraction. Why? Because they know their terrain (network topology) better than anyone else, and they understand the behavior of the network entrusted to their defense. They are also the only ones who face the adversary, and they know the effects of cyber attacks as experienced from the inside. An outsider, regardless of good intentions, lacks the empirical realities of such effects. These internal cyber defenders are also the ones who understand best how to coordinate the mitigating and defensive COA. Generic cyber defense strategies may certainly contribute to mitigate cyber attacks, but ad hoc cyber defense strategies are always the best COA.

CYRE is harsh, is ugly, is merciless. CYRE is based on cyber code, which can be written to build or destroy, to enlighten or to obfuscate. The cyber code is a creation of man, and as such, is tailored by the motivation and the purpose in the mind of its creator. This is the objective side of CYRE, and it may or may not correspond to any personal subjective CYRE. The cyber arena is a jungle, and you cannot expect to prance around this jungle, without a care or concern, and assume you won't become a victim of the myriad of cyber predators waiting for an unaware victim, or one that think: this won't happen to me. Because the cyber adversary is a human being, only another human being can thwart the attack. A cyber adversarial encounter is primarily a battle of brains; the presence of procedures is only ancillary.

Chapter 17. Becoming a Cyber target

Is anyone of us susceptible to becoming a cyber target? Statistically, the answer is yes, simply because the majority of cyber targets are targets of opportunity. Therefore, if our cyber behavior creates the conditions for anyone of us to become that target of opportunity, we will become a cyber victim, once the cyber predator identifies us as a target of opportunity.

Becoming a cyber victim requires primarily two conditions: we personally and directly create the conditions to be identified as a target of opportunity, or we may collectively become a target of opportunity when our binary data becomes part of a group target, as in the case of a compromised database containing our personal data.

Whether by choice or by design, our lives are immersed in the global cyber environment. Our identity (ID), with all the particulars defining us as a person, resides on a digital record, contained within a database, and consequently, on a discoverable status. When someone creates, edits, add, or deletes a data entity on our digital record, our lives are impacted by any changes on our digital record, and propagated among all the cyber devices used for the storage and processing of digital records.

We have all suffered at one time or another the consequences of erroneous data entry inserted into our digital records. Perhaps the biggest threat factor to our persona in CYRE is the unconditional trust we place on digital records, thinking and assuming, as an expression of our collective naiveté, that cyber devices are infallible. We have, as a society at large, transferred our trust from the individual to the digital record representing that individual. This shift has migrated the trust factor from a human being to a digital compilation of data, thus ascribing a greater degree of trust to an impersonal record. This shift in the trust factor has placed humans at the mercy of digital records. This shift is unnatural and misplaced; the essence of the individual is not transferred to the digital record. This latter is simply a flat digital compilation of data, in a similar manner in which a photograph is a flat representation of the external physical appearance of an individual, incapable of capturing the multidimensional reality of the photographed individual. Conversely, the multifaceted reality of the personal essence cannot be captured on a digital record, which is subject to manipulation and errors.

The shift in the trust factor from the individual to a digital record is a dangerous migration. How many of us have already been negatively impacted by errors introduced in our digital record? How many of us have been confronted by a person accessing our digital record and telling us we are wrong, simply "because our computer says so"? Computers do not create or edit our digital records; humans do, and humans are consistently and unavoidably prone to errors. So, when a discrepancy in our digital record is detected, the error originated with the person who created or edited our digital record.

This fact of life requires that we confirm the data entered into our digital record is accurate. No one else but us cares about the accuracy of data entry into our records. We should not trust implicitly the person collecting and entering our data into digital records; we should instead require confirmation at the moment of submitting our personal data. We should treasure the integrity, accuracy and security of our personal data, and take every precaution to ensure that only authorized individuals have access to our digital records. Data manipulation in our records by unauthorized individuals will lead to disastrous results.

Cyber Reality

Regretfully, we often facilitate intrusions into our personable Identification information (PII). How, may you ask? By submitting too much PII. For instance, you fill out a form at your doctor's office, containing many fields that are not truly relevant, and quite often too intrusive. Submit only the minimum data required. It is also very common to create online accounts, such as web-based emails accounts, and make the error of submitting private information, quite often containing information on our full name (first.last@whatever.com). Since emails are forwarded and circulated on a global scale, we are actually contributing to the disclosure of an important element regarding our personal identity, at a large scale, to people we do not even know.

Furthermore, we are facilitating the opportunity to become the target of opportunity for an adversarial cyber activity. Remember the database we briefly mentioned at the beginning of this chapter? If I disclose my full name on my web-based email, then a malicious cyber actor may intrude into that database (DB) containing my PII, and now I have become *ipso facto* a cyber target of opportunity. Now I am exposed to suffer serious consequences in my professional, financial, and personal life. If my PII in the compromised DB includes my SSN, now I am a prime target for fraud and identity theft. This is another dark aspect of our CYRE; becoming a cyber target for unscrupulous cyber predators.

Is there a way to mitigate this problem? Very succinctly, yes. Reduce the amount of PII you disclose, and use a pseudonym in your email accounts. If you have already disclosed too much PII, simply increase your degree of vigilance. Do you already have an email account with your full name? Terminate it, and create another one with a pseudonym. Is this too radical a solution? Do you want to remain as a global cyber target? An remember, once you become a cyber target, you are also contributing to the endangerment of those who have a cyber connection with you. SNS are the hunting ground for cyber predators, and once you become identified as a cyber target to a cyber predator, your whole circle of SNS contacts become exposed as well, and identified as additional potential targets.

A cyber predator may easily impersonate a cyber victim, and approach the entire SNS circle of contacts of that victim. If the cyber predator intrudes into a DB containing your PII, the same DB may also contain PII information for your family members and those belonging to your SNS circle of contacts. Therefore, when you facilitate becoming exposed as a target to a cyber predator, you are also facilitating the exposure of your circle of contacts. The main action to initiate mitigation and control damage toward protecting your personal cyber identity is to reduce your cyber footprint. If you have already contribute to a large cyber footprint, then increase your degree of vigilance and remain alert.

Perhaps the biggest threat factor to our persona in CYRE is the unconditional trust we place on digital records, thinking and assuming, as an expression of our collective naiveté, that cyber devices are infallible. We have, as a society at large, transferred our trust from the individual to the digital record representing that individual. This shift has migrated the trust factor from a human being to a digital compilation of data, thus ascribing a greater degree of trust to an impersonal record

The excessive amount of PII already disclosed is

Cyber Reality

irreversible, and cannot be reduced; it is already stored, disseminated and replicated many times over in the global cyber environment. Thus, vigilance is the only feasible course of action. If you cyber footprint is small, keep it that way, and provide information only on a need-to-know basis, and only to authorized people. A small cyber footprint will also decrease the risk of becoming a cyber target, and minimize the exposure rate of your circle of cyber contacts.

When traveling or relocating, are you aware of the identity of the ISP you are using? When flying, and connecting to in-flight Wi-Fi service, are you aware who is the organization providing that service? Conversely, are you aware of what type of cyber systems is the Wi-Fi provider using to process and forward your cyber data? Is the Wi-Fi provider, or the cyber system provider (if different) respectful of the laws protecting your privacy in the US? If either the Wi-Fi provider or the systems provider are not respectful of the US laws protecting your privacy, how are they managing your cyber data traversing through their cyber systems? If you are using this in-flight service when traveling, are your login credentials kept private and protected from disclosure?

Our appetite for persistent online coverage may easily over rule our sense of vigilance and caution, and we may divulge an excessive amount of information potentially leading to becoming single out as a cyber target. If you have no information available about the ISP offering its services when you are traveling or relocating, then be very spartan regarding the amount of information you share through the available ISP.

When relocating to OCONUS,[204] increase your vigilance regarding your cyber profile activity, and decrease your cyber footprint. There are very reliable encryption tools available, and they provide an additional layer of security for your

---

204 Outside Continental US (OCONUS)

cyber interaction with your circle of contacts. If you do not feel comfortable using encryption, then you may choose a very cryptic manner of language, designed to obfuscate the meaning of your dialogue to all casual observers or listeners that are not the intended recipients.

If you feel comfortable with encryption, but you are apprehensive about the commercially available encryption tools, then you may consider creating your own personalized encryption tool, with readily available Java code that allows you to write your own encryption device. They may offer only a succinct level of communication, but these Java-based solutions are tailored for the exclusive use of the originator and the intended recipient. They are completely unintelligent to any one else.

When residing in the US, and contracting the services of a US-based ISP, you have a pretty good chance of having your online data protected by the US privacy laws. However, when traveling or relocating, do not assume you have the same protection. Conversely, do not assume that when using in-flight Wi-Fi services in the US that your cyber data is protected. These Wi-Fi services contracted by US carriers do not necessarily employ the services or the cyber equipment of US providers. Because of the complexity of the supply chain, the Wi-Fi service, and the cyber equipment, may be provided by foreign companies. Thus, when using Wi-Fi services on an American carrier, be austere on the amount and type of data you share through the in-flight Wi-Fi service.

There was a very insightful article published in 2011,[205] addressing the security vulnerabilities associated with accessing in-flight Wi-Fi services. These type of articles are

---

205 Marc Weber Tobias, Insecure WiFi At 30,000 Feet, 6/27/2011, http://www.forbes.com/sites/marcwebertobias/2011/06/27/insecure-wifi-at-30000-feet/

very rare, simply because the majority of articles addressing in-flight Wi-Fi services are written by enthusiastic users who are bewitched by the conveniences of such service, but blissfully incognizant to the dangers of such service.

This article details the actions undertaken by a penetration specialist who connected to the Gogo Wi-Fi on board an American carrier. This user conducted port-scanning on the wireless router on board the aircraft, and gathered information on all the passengers logged into the service. He also connected to a server in Asia, suspected of being a hacker haven, and examine its local configuration. The author of the article contacted Gogo and presented to them a detailed request for technical information on their cyber environment security, and that of their clients. Gogo refused to provide any information.

The activities performed by the penetration specialist can be replicated by any other passenger behaving as a malicious cyber actor. He or she could be scanning the Wi-Fi traffic, capturing passengers information, injecting code onto their computers, or launching a rogue access point to capture confidential data.

The bottom line? Reduce your cyber footprint whenever possible. Else, be spartan in your written communication, or cryptic, or both. If you feel comfortable using encryption tools, do so. Following this cautious modus operandi will reduce the possibilities of becoming a cyber target of opportunity for cyber predators. It will also enhance the protection factor for your circle of cyber contacts.

Chapter 18. Facing the Truth of CYRE

Why do we in general take the path of interacting with CYRE at a casual level? Why do we really love to engage with the CYRE dimension as if it were an amusement park? Is it because of lack of understanding of the awesome power ingrained in CYRE? Definitely not, since we are being made aware of the amazing capabilities of CYRE on a minute basis. We are reminded of the immense power of this man-made reality every single second of our modern existence. We wake up surrounded in this reality, we live our personal and professional lives immersed in the CYRE, we think, interact, and we go back to sleep cradled in CYRE. Is it because is so pervasive, with a quasi-omnipotent quality that we take it so casually, just as we do with the air that we breathe, so casually as well?

CYRE apparent omnipresence is also remarkable fragile, and definitely not perennial. CYRE is dependent on the availability and survivability of the Internet, and the Internet is dependent on the survivability of the Internet infrastructure, and the Internet infrastructure is dependent on the electronic power grid, and the power grid is extremely susceptible to astronomical disturbances, already summarized by this author on a previous book.[206]

Retroactively, we can certainly state that we did not experienced a catastrophic solar coronal mass ejection (CME) episode during 2013, but the cyclical nature of these astronomical events renders them an impending reality. The disruptive and destructive power of these violent explosions of solar particles cause significant disruptions on our planet's geomagnetic protective field, affecting and eventually destroying electronic equipment.

---

206 Giannelli, chapter 14

And then, right at the beginning of 2014, on January 9 the Sun released a CME. [207] The human population is protected from the harmful radiation of these violent solar flares by the Earth's atmosphere, but the same cannot be said of our electronic devices. These CMEs, depending on their intensity, have the power to disrupt the atmospheric layer where GPS and communications signals travel, and when more severe, the electronic equipment on the ground as well. This particular January 9 event was rated as an X1.2-class[208] flare, triggering only an aurora borealis phenomenon. Subsequently, on January 28,[209] NASA's IRIS space telescope[210] allowed scientists to witness the eruption of another solar flare, labeled as an M-class[211] flare.

These observations of CMEs may represent the prelude to upcoming solar activity, generating events with enough intensity as to affect global economies, increasingly dependent of cyber programs and services, and thus, increasingly vulnerable to the cyclical nature of solar phenomena. Intense CMEs will disrupt power grids, and consequently civilian and military cyber operations in general, and communication and navigation systems so dependent on GPS signals in particular.

The objective CYRE is ephemeral and fragile. How will the

---

[207] NASA, Weak CME Arrives and Sparks Aurora, January 10, 2014, http://www.nasa.gov/content/goddard/sun-unleashes-first-x-class-flare-of-2014/#.UwjXUtsy3Cl

[208] The level of intensity is signified by the numeral value following the X designation, progressing from X1, to a double-strength X2, to a triple-strength X3, and so on.

[209] ScienceDaily, NASA's IRIS spots its largest solar flare, February 21, 2014, http://www.sciencedaily.com/releases/2014/02/140221153146.htm

[210] NASA, IRIS: Interface Region Imaging Spectograph, http://iris.gsfc.nasa.gov/

[211] The M-class flare is categorized as the second strongest class flare after the X-class

global population adjust their subjective CYRE when a catastrophic CME dawns on us? Dependence on a particular technology develops habits and expectations, and set the practitioners on a particularly debilitating status when such expectations are not met. What will you do when you no longer can reach your favorite SNS, and tell the world who you are, and what you're doing? What will the enterprise do when the access to data, and the production and processing of data is a reality no longer available? What will the economical impact be when marketing activity and transactions are interrupted by the unavailability of the cyber infrastructure?

Will a powerful CME event disrupt the cyber infrastructure? The historical and scientific empirical data we have accumulated on CME, in addition to the related electromagnetic pulse (EMP)[212] phenomenon, leaves no doubts about the crippling effect generated by them. Thus, the labeling of the objective CYRE as an ephemeral and fragile reality. Enjoy it while it lasts, because it won't be available for a long time, only until the occurrence of the next powerful CME wielded by our central star, at a global level, or by the next disruptive EMP, at a regional level.

After the crippling effect of the inevitable upcoming CME, if reaching to the level of global effects, will we be prepared to cope with the absence of CYRE? What would it be the scope and criticality of a significant CME of global proportions? The intent of this chapter is not to induce a panic wave of uncertainty, but rather a mature and proactive motivation to prepare global, regional, and personal contingency plans.

There are a number of USG organizations dedicated to monitor the cycle of solar disruptive events that may affect

---

212 See Giannelli, chapter 14, for a brief account on the EMP phenomenon.

directly or indirectly the electro-infrastructure in our country. The National Oceanic and Atmospheric Agency (NOAA) maintains a constant monitoring operation on solar activity, and maintain records of every observable solar event that may affect our planet. Solar disturbances include solar storms and solar flares, because they create adverse effects on the magnetosphere of the Earth. Solar flares are sometimes followed by CMEs, that may reach a diverse range of intensity. For instance, NOAA's monitoring activity reports include, as of 18 March 2014, a total of 54 solar events for the current year. Of these, 28 generated CMEs, with 25 CMEs affecting the Earth, directly or indirectly.[213]

We do not have yet a clear understanding of CMEs and their effect on Earth. The data we currently have does not reveal a clear pattern. For example, we know of cases where CMEs occasionally damaged satellites and disrupted terrestrial power grids. In 1989 Quebec suffered a nine hours power outage affecting six million people, and in 2003 South Africa experienced damage on their power grid main transformers. Yet, in July 2012 we made global preparations to confront a series of three CME targeting Earth, but upon arrival they simply produced spectacular aurorae without causing any outages. Some CMEs cause disruptions, while others of similar intensity do not, and researchers are beginning to work on some models that may allow us to discriminate between one kind from the other.[214]

Any CME targeting the Earth may have catastrophic effects on our planet's magnetosphere, because a CME occurring under the required conditions has the power to produce not only regional, but global and disastrous consequences, requiring an extended amount of time before we could

---

213 NOAA's Space Weather Prediction Center, http://swpc.noaa.gov
214 Saswato Das, Short-Circuiting Civilization: Predicting the Disruptive Potential of a Solar Storm Is More Art Than Science, Aug 23, 2012, http://www.scientificamerican.com/article/short-circuiting-civilization/

Cyber Reality

implement a recovery plan, in the event of having such a contingency recovery plan. Do we have one?

Considering the interdependencies among nations regarding power generation, would a local or regional contingency plan suffice? And if a joint global plan would be required, are we in the process of establishing one? Is it feasible, under the perennial adversarial global relationships, to conceive and create such a joint contingency plan?

When the arrival of the next powerful and disruptive CME occurs, the intensity of this CME will damage the main transformers of local and regional power grids, and it will be difficult to arrange for an expeditious replacement. These main high-capacity transformers require a year or more to manufacture, and even if we still were to have the knowledge on the manufacturing process, will we have the required equipment to build the replacement transformers? If we are being affected by an extended power outage, where will we obtain the power source to enable the manufacturing process for the replacement units?

Are you now getting the picture of your reality without CYRE? How would we sustain our contemporary complacency on assuming that CYRE will always be available? How would we manage without the availability of IoE?[215] What will you do without the ubiquitous wireless networking availability? How will we coordinate our national tasks, our national security missions, our interdependent global economy?

When it comes to CMEs we need to abandon a provincial outlook and think in global terms, since the global interaction in all aspects of human activities is now intrinsically connected. Global economies are inevitably vulnerable to the dynamics of solar activity, and multi-level human affairs

---

215 See chapter 12 above

are exposed and affected by CMEs capable of disrupting power grid infrastructures, and interfering with civilian and military airborne and ground communications.[216]

The tasks of monitoring and forecasting solar disrupting activity are entrusted to NOAA, as part of our national priorities. NOAA's Space Weather Prediction Center (SWPC) is the official source of space weather forecasts, alerts, and warnings. On a not-to-distant day, SWPC will issue an alert that a large sunspot unleashed a solar flare, followed by a significant CME targeting Earth, and three days after ...

Are you there...? Can you hear me now?

Considering the interdependencies among nations regarding power generation, would a local or regional contingency plan suffice? And if a joint global plan would be required, are we in the process of establishing one? Is it feasible, under the perennial adversarial global relationships, to conceive and create such a joint contingency plan?

The truth about CYRE is that it is an ephemeral, man-created reality, existing in the cyber binary code, that requires cyber systems built and configured to read, interpret, and process such binary code. These systems are dependent on the availability of an interconnected network sustained by a power grid infrastructure. Therefore, the CYRE is dependent on the power grid that sustains it.

Your personal CYRE, based on your personal concept of

---

216 http://www.noaa.gov/features/01_economic/spaceweather_3.html

CYRE, extrapolated from whatever factual concepts or misconceptions you weaved from your personal and professional experiences in adopting CYRE, are equally dependent on the availability of the power grid. Any disruption to the power grid will affect your CYRE, and if that disruption is beyond human control, so it's your CYRE. Are you prepared to live in the absence of CYRE? Do you have a contingency plan? Do you?

Chapter 19. A Taste of CYRE

At the time[217] of compiling the index for this book there was an extremely interesting and well publicized data breach event affecting a notorious and important US financial institution. This very current case of data breach offers a timely topic to close this book by providing invaluable analytical opportunities to discuss the concepts presented in this book. Based exclusively on the views presented in the media, let's conduct a cursory review of the several principles outlined and discussed in this book. Why? Because we will see many of the discussed aspects of CYRE coming to life in this case. Assumptions, misconceptions, flaws, omissions, myopic views, distorted views, incognizant views, and above all, an endemic detachment from the CYRE principles. Ingrained in the views and comments reported by the media we see society at large and organizations in particular still attempting to force CYRE into unrealistic and obsolete models.

Taking information from assorted media sources, we'll analyze the views and expectations represented on a few media articles recently published about this current event, labeled as a disconcerting data breach. Let us focus on three primary items.

1. A massive data breach affecting a reported 83 million households and businesses accounts.[218] During the course of the breach, spanning several months, the cyber intruders maintained a persistent access to the financial institution's

---

217 October 2014
218 Danielle Douglas-Gabriel, The Washington Post, JPMorgan breach raises alarm about safety of financial system, October 6, 2014
http://www.washingtonpost.com/business/economy/jpmorgan-breach-raises-alarm-about-safety-of-financial-system/2014/10/06/f9f9dff0-4d80-11e4-babe-e91da079cb8a_story.html

compromised network, returning at least five times[219] over a period exceeding three months. There have been no explanations as to the reasons why the cyber predators were able to obtain and maintain persistent access to the network of this financial institution. There is no explanation as to why the cyber intrusion went undetected for such a long period of time. There is also a peculiar omission in all the accounts and comments reported by the media; there are no interviews to qualified cyber operators. All the comments come from executives, who ponder how this data breach could have occurred to a financial institutions ascribed by many of the commentaries as a bastion of cyber security.[220] Is this expectation a realistic one? Or rather represents an unfulfilled aspiration based on distorted views on CYRE?

2. In one of the numerous blogs published online is where we can find what it is perhaps the most shallow and disingenuous comment, born of misconception and incognizance regarding the essence of cyber. The managing director of a professional services advisory firm comments that the data breach event is incomprehensible, given the fact that the Chief Information Security Officer (CISO) responsible for the cyber protection of the compromised financial giant is a person he considers as someone with great cyber experience. This managing director adds that the Security Officer was the "gentleman who was at the forefront of defending attacks against the U.S. Air Force."[221]

Let's analyze the logical and empirical flaws in this comment. First, the CISO had only recently assumed this position,[222]

---

219 Ibid
220 Ibid
221 Hugh Son and Michael Riley, Bloomberg, JPMorgan Had Exodus of Tech Talent Before Hacker Breach, Sep 4, 2014, http://www.bloomberg.com/news/2014-09-05/jpmorgan-had-exodus-of-tech-talent-before-hacker-breach.html
222 Ibid

and it's irrational to expect that his previous employment experience will automatically empower him to thwart a cyber attack at the very beginning of his new employment. The cyber security flaws facilitating the data breach were pre-existing to his employment.

Second, a military commander functions at the level of an executive, not at the level of an experienced and technological expert in cyber security matters. The alleged role of being "at the forefront of defending attacks" is a gross misunderstanding of a commander's role. He functions as an executive leader, not as a cyber operator. His lack of specialized technical cyber knowledge is clearly evident in his writing,[223] where his lexicon is dominated by generalizations, lacking indicators of cyber technical terminology and concepts. A commander depends on cyber SMEs for the fulfillment of the defensive mission, while he himself is not a cyber SME.

Even though this CISO published a book on cyber matters, his writing does not address the technical aspect of the cyber dimension. He embraces the same dislocated misconception of cyber by equating it to electromagnetism and light, erroneously associating these two as encompassing "the principal physical laws governing cyberspace."[224] As I have pointed out on several occasions,[225] EMS belongs to the sphere of radio frequencies and waves, while light manifests itself in both waves and particles form. Cyber is neither waves nor particles; cyber is essentially binary code, nothing more, nothing less. Electromagnetism and light are simply conveyances for the binary code, but they are not part of the binary code.

---

[223] Gregory Rattray, Strategic Warfare in Cyberspace, Cambridge, MA, MIT Press, 200, chapter 10,
http://ctnsp.dodlive.mil/files/2014/03/Cyberpower-I-Chap-10.pdf
[224] Ibid
[225] In this book, and my previous one, Giannelli, The Cyber Equalizer

Cyber Reality

The CISO's writings display another area of incognizance regarding the essence of cyber. At this point I have to reiterate once more that telecommunications precede cyber, but telecommunications are not cyber. The CISO shares the same distorted view about cyber as a myriad of other analysts attempting to force the cylindrical peg of cyber into the cubical hole of EMS and telecommunications. This misconception is clear when he draws analogies from the US initiative during both World Wars of taking control of the primary US telephone provider.[226] Not everything in the electronic environment is cyber. On the other hand, everything cyber is dependent on an electronic environment.

The dislocated concept of cyber is so entrenched in this CISO that he adds: "The extension of conflict to cyberspace began as early as the Crimean War when the telegraph was used to transmit intelligence reports and to command widely dispersed forces."[227] This statement is proof of both a crass lack of understanding on the essence of cyber, and an gross anachronism as well. The Crimean conflict dates back to the 1853 – 1856 period, a time when binary code had not come into existence yet.[228]

I'd like to close this section on the CISO by highlighting what is perhaps his most radical departure from the proper understanding of cyber. He pontificates that "the Internet was not designed to help track the origin of activity."[229] All it takes is a rudimentary knowledge of the *de facto* Internet TCP/IP suite of protocols to prove that it constitutes the backbone of our Internet traffic, and it was indeed specifically designed to track the origin of network datagrams. The very

---

226 Gregory Rattray, Strategic Warfare in Cyberspace, http://ctnsp.dodlive.mil/files/2014/03/Cyberpower-I-Chap-10.pdf
227 Ibid
228 See previous Chapter 9. Abridged History of Computing
229 Gregory Rattray, Strategic Warfare in Cyberspace, http://ctnsp.dodlive.mil/files/2014/03/Cyberpower-I-Chap-10.pdf

operational foundation of the IP protocol is to have an IP header with the source IP address and he destination IP address.[230] What we also have to confront in the cyber dimension is a series of obfuscating techniques the cyber predators use for their advantage, making the tracing of the origin very difficult to locate, but not impossible.

This cursory review of the CISO's views, and a sizable amount of other misconceptions found in his writings reveal his non-technical formation.[231] So, why is it that the community of concerned individuals highlighted in the media reports think that this CISO could avert the data breach? He is not a qualified and experienced cyber SME. He is a leader, a strategist and analyst.

3. This takes us to the last comment in this case study; the expectations of cyber security capabilities on the network of the compromised financial institution. The comments in the media state that the victimized institution had an alleged yearly spending of $250 million invested in its defenses prior to the data breach.[232] Despite this investment, the cyber perpetrators gained access to the victim's network in June 2014, and maintained their access until the discovery of the intrusion in August 2014. At this point we also have to highlight the fact that there are no comments regarding any cyber forensic evidence documenting the cyber predators' access has been conclusively terminated. Therefore, this will remain an open question.

---

230 Request for Comments (RFC) 1180, January 1991 http://tools.ietf.org/html/rfc1180
231 This CISO also authored a PowerPoint presentation in 2013, consisting of a potpourri of disconnected pictures and bullets, lacking any insight into the technical dimension of cyber. See The Evolution of Cyber Threats and Cyber Threat Intelligence, Greg Rattray CEO, Delta Risk LLC, 22 March 2013
232 Danielle Douglas-Gabriel, The Washington Post , JPMorgan breach raises alarm about safety of financial system, October 6, 2014

None of the media reports states how the intrusion was detected, and there is no explanation how the budgeted $250 million were distributed to satisfy the mission of cyber defense. Was this budget allocated to the acquisition of automated cyber detection systems? And if so, was the budget also covering the employment of qualified and experienced cyber SMEs? The fact that the intrusion was discovered after three moths may clearly indicated that the cyber defense budget may have covered only the operation of autonomous cyber defense systems, without the supervision of qualified cyber defenders. Why? Because when you have a small contingent of qualified cyber defenders, experienced in deploying and operating a proactive cyber defense program, it does not take three months to uncover a cyber intrusion. How do I know this? Because I was the lead of a cyber defense team for over 8 years, and it usually took us only hours, and sometimes only minutes, to discover an incipient cyber intrusion. If you have the proper equipment, and more importantly, the proper personnel, dedicated to actively monitoring the status of the network entrusted to your care, you can certainly detect the network traffic anomalies representing the telltale of a cyber intrusion.

I participated as a cyber security consultant on a case where anomalous cyber traffic was observed. By the time I joined this case, the group dedicated to ascertain the root cause of this network traffic anomaly had already spent 22 days researching the root cause, and attempting to discontinue the anomalous traffic, unsuccessfully. I requested the logs corresponding to the peripheral networking devices involved in this issue, and after a one-hour forensic analysis of these logs I directed the network administrators to implement the corrective COA. The team who spent 22 days without finding the root cause were not qualified to perform forensic analysis of the affected systems' logs.

Cyber Reality

So, how were the $250 million allocated in the cyber security budget? Were the executives in charge of disbursing the funds knowledgeable on how to allocate the funds in the most effective manner to achieve a strong cyber security posture? Did they know what to buy with the $250 million? To say that an institution spends $250 million in cyber security has no meaning in itself. The only meaningful manner of disbursing the funds is to consult with cyber SMEs who knows how to allocate this budget, and a significant portion of this budget should be invested in qualified and experienced cyber security SMEs. Cyber conflicts are fought with the brains of qualified cyber SMEs, not with a collection of autonomous cyber devices operating without the supervision of these qualified and experienced cyber SMEs.

Cyber is neither waves nor particles; cyber is essentially binary code, nothing more, nothing less. Electromagnetism and light are simply conveyances for the binary code, but they are not part of the binary code. Telecommunications precede cyber, but telecommunications are not cyber. Not everything in the electronic environment is cyber. On the other hand, everything cyber is dependent on an electronic environment.

The overall media comments emphasize that Wall Street's financial institutions have a critical place in the US economy, and this is an incontrovertible truth. Therefore, we must dedicate the best cyber security resources to this task, but the decision of what constitutes the best cannot be reached without consulting with the best cyber security SMEs. Were such SMEs consulted in designing and deploying the cyber security model for the critical financial institution that fell victim of cyber predators? Were such SMEs employed to actively defend the cyber security operation in this institution? The indirect evidence that can be extrapolated from the choir of mourners and their commiserating comments indicate the answer is no.

Lessons learn. When the expectations for cyber security are based on a distorted and dislocated concept of CYRE, don't expect victory, but defeat instead. Executives cannot design or guarantee an effective cyber security posture. Autonomous cyber security equipment cannot do it either. There is no such thing as an impervious cyber security model, but when it comes to protect a critical asset we have to listen to the professional advice of true cyber defenders. They are a special breed, and they are not cheap, because of their specialized knowledge. How many truly qualified and experienced such SMEs were in the employ of the targeted institution? How many of them were interviewed by the media, to provide an accurate forensic assessment of the data breach? How many of them are involved in the mitigation and recovery process following the discovery of the data breach?

We need to stop lamenting our losses, stop depending on dislocated and distorted cyber expectations, abandon unjustified personal opinions about CYRE,[233] and start depending on the advice and professional expertise of qualified and experienced cyber SMEs. No one who depends on traditional theories and strategies can design and maintain effective cyber defenses. Only those who confront the cyber adversary at the level of network raw data and binary code are able to do so.

The cyber dimension cannot be understood by analogies, because there are no historical antecedents for the cyber

---

[233] Consult again Chapter 11. The perception of reality

dimension. Those who rely on historical theories and strategies cannot fathom the depth and uniqueness of the cyber dimension. The cyber dimension is a reality unto itself, without paragon, without parallel. CYRE can only be understood through the essential principles of CYRE.

Epilogue

This author offers this book as a contribution to the goal of obtaining a more comprehensive and balanced understanding of cyber reality from an ontological and epistemological perspective. This concept has somehow been neglected in the plethora of literature dedicated to the presentation of cyber matters.

Why do we as a modern society hold a rather casual and superficial knowledge of a science and technology that occupies a central and critical role in our global society? If we invest time in gaining a proper understanding on how we acquire knowledge about cyber reality, the ubiquitous dimension that encompasses every aspect of our lives, at both the private and public level, perhaps we may reach a more mature and balanced conceptualization of CYRE.

This book brings ontology and epistemology into the process of understanding CYRE. The hope is that in fusing these two aspects with our conceptualization of CYRE, we may become enlightened regarding the background processes leading to the rather overly simplistic and casual popular concepts pervading the immature global perception of CYRE.

Until the end of the cyber age, CYRE dominates both our private and public lives, at the national and international levels. CYRE has indeed transformed our world into a village, where everyone and everything exist in a very close proximity. CYRE has demolished the traditional boundaries, and we now co-exist in a very small neighborhood of interconnected entities. The concept of a real far, far away environment is no longer part of our global reality. Everything is very close when it comes to the digital age.

Cyber Reality

Nothing is out of reach for those who know how to get there.

This author hopes this book is both an awakening and an enlightening experience, leading the reader to realized that CYRE is both a wonderful and ominous reality at the same time, coexisting in superposition. Rejoice at the wondrous opportunities offered to you by CYRE, but be cautious when confronting the dangerous aspects of CYRE as well. The allure to over expose yourself before the global community is powerful, and even rewarding, when given the opportunity to tell the world who you are. Yet, at the same time you increase your visibility, the bigger target you become to a throng of cyber predators.

Perhaps an analogy may help to convey this advice in a more palatable way. The pleasure of a good wine ends when you become inebriated. Therefore, the mature and balanced option is to enjoy the pleasure offered by an exquisite wine,[234] enjoying its taste and aroma, and the accompanying sensations, while exercising the proactive caution of knowing when to avoid the over indulgence of becoming inebriated.

---

[234] Of course I am referring to an authentic Italian wine. There is no other kind meeting the requirements presented in this scenario, (smile). Perhaps an Archeo Sicilia from the Nero D'avola grape family, to enjoy with a red sauce dish and grilled meat?

Alphabetical Index

a battle of brains...199
a statement of intent..124
abaci..99
absence of CYRE..213
absolute knowledge...15
abstract nature of CYRE...137
abstract reality..136
abstraction methodology...88
abused speech crutch..86
accountability and penalties..126
accuracy of data entry...201
accurate forensic assessment..221
Ada Lovelace...85
adjacent AS border routers..83
adversarial agents...44
adversarial cyber activity..27
age of innocence...82
airborne and ground communications.................................212
algorithm..85
allure of the IoE devices..169
alter ego..42
alternate cyber persona...42
alternate pseudo identity..42
amateurish degree of technical skills....................................82
Amdahl's law..106
American network pioneers...174
amplification of cyber security risks....................................162
an impending reality...207
analytical and proactive cyber defense..................................30
Analytical Engine..85
Android...168
anonymous dimension...39
antipodean..39
antithetical realities..11

225

ARPANET	173
articulation of a derived reality	131
artificial dichotomy	123
AS	78
ASCI Red	105
ASN	78
assembler	89
assembly language	52
astronomical disturbances	207
asymmetrical	28
attack environment	193
aurora borealis	208
authentication servers	84
autobiographical facts	135
automated cyber detection systems	219
automated detection tools	29
autonomous developers	166
availability and survivability of the Internet	207
bandwidth availability	155
behavioral choice	41
Big Data	145
Big data analysis	142
Big Data phenomenon	150
binary format	89
binary language	27
binary nomenclature	88
Bluetooth technology	170
border gateways	181
border routers	83
botnets	43
BRIC coalition	178
careless handling of sensitive data	68
careless SNS contacts	45
Cartesian dictum	139
cautious modus operandi	206
CDC	194
Chaos Theory	151

circle of SNS contacts	203
circuit-switched technology	173
civilian and military cyber operations	208
classical computing	26
CLI	92
cloak of infallibility	139
cloak of invisibility	35
cloud provider	159
CME	207
code execution	131
cognition	8
cognitive background	134
cognitive process	130
cognizant interface	122
cohesive and logical instructions	87
collective anonymity	35
collective naiveté	200
collusion between ITU and BRIC	178
command line	92
commitment to caution	56
common deficiency in human communication	86
common routing policy	78
compiler	89
complex data sets	151
compromised database	200
compulsory appetite for disclosure	35
computing cluster platform	101
computing device's native language	88
computing pioneer Charles Babbage	85
conceptualization of the surrounding reality	132
conceptualized entity	131
confidentiality and integrity of PII	127
configuration settings	36
connectivity growth	156
conscious entity	137
conscious observer	138
contingency plans	209

cost of a floating point operation...113
cost of moving data off-chip..113
costly research data...194
counterfeit cyber devices..192
COWEP..57
CPU..89
critical..31
critical infrastructures..31
cryptanalysis operation..60
cryptographic keys...154
current modern social structure...123
cyber "schizophrenia"..40
cyber activity..36
cyber alter ego..45
Cyber ambiguity...32
Cyber anonymity..84
cyber attack methodology...27
cyber behavior..200
cyber business..38
cyber code weaponization...50
cyber collection and espionage..194
cyber conflict..27
cyber creativity...131
cyber criminals...43
cyber defense strategy...29
cyber definitions...31
cyber developer..131
cyber environment...131
cyber forensic evidence...218
cyber global repository..141
cyber identity..34
cyber identity theft...37
cyber irresponsibility...41
cyber is essentially binary code..216
cyber malfunctions...31
cyber memory...88
cyber misconceptions...10

cyber neophytes	77
cyber persona	76
cyber presence	34
cyber processor	87
cyber profiling	84
cyber programmer	87
Cyber programming	85
cyber protection	54
cyber prowess	27
cyber security	56
cyber security breach	124
cyber security experts	197
cyber security policies and procedures	126
cyber society	34
cyber spectrum is nonlinear	55
cyber target	200
cyber terrorism	50
CYRE	10
CYRE cacophony	139
CYRE knowledge	28
CYRE principles	214
DARPA	173
data breach	214
data entity	200
data segments	77
decompose the abstract concept	87
decrypted data	60
dedicated circuit	173
deductive reasoning	66
degree of separation	44
degree of vigilance	202
demagogic and hyperbolic statements	196
democratization of opinions	140
democratization of thoughts	47
dependencies on cyber technologies	192
dependency on cyber code	54
destination IP address	218

DID-affected entities..39
DID-affected generation...49
digital overexposure...75
digital records..122
digital records exposure..125
digital replication...151
dimension of interpretation...82
disaster recovery..159
disclosure of sensitive information..45
discoverable status..200
dislocated misconception of cyber...216
disruptive technology..162
dissemination agent for spyware...43
dissociative identity disorder..41
distorted sense of CYRE...194
distorted sense of ethics...66
distorted sense of reality..39
distorted view of CYRE...48
DNS..176
DNS authoritative root zone system..183
DoE...105
dualistic identity..39
effective cyber defense strategy...198
Egalitarianism...140
Einstein...130
electro-infrastructure...210
electronic power grid..207
elimination of global control...175
embedded cyber code..191
embedded cyber systems...191
embedded systems..167
EMP..209
empirical awareness..135
enabling terrorism...59
end-to-end security scheme...154
entangled sub-atomic particles..17
entertainment sandbox..56

entity stored as a digital record	123
ephemeral and fragile reality	209
ephemeral condition	17
epistemic process of knowledge	133
epistemology	132
erroneous data entry	200
Espionage Act	64
espionage laws	72
essence of cyber reality	134
Ethernet technology	174
Ethical behaviors	44
ethical model ruled by relativity	44
exascale	107
expectation of privacy	141
exponentially expanded network	163
exponentially larger consequence	55
external and sensory stimuli	131
extrapolation	16
false dichotomy	139
filtering and blocking technologies	180
FIPS 140-2 certified encryption module	128
first computer network	172
floating-point operations	102
FLoating-point Operations Per Second	109
FLOPS	102
foreign adversaries	193
foreign intelligence entities	192
forerunner of the Internet	173
free and uncensored Internet	176
FTC	166
FTP	159
fused intelligence	83
GAC	177
geomagnetic protective field	207
geospatial data	146
GFLOPS level	102
Gigapolis	46

global and decentralized Internet..176
global connectivity..155
global cyber environment..200
global Internet village..42
global platform for data leaks...43
global supply chain..190
globalization of manufacturing..189
GPS...208
gross anachronism...217
GSA..127
GUI..92
healthy degree of caution..46
heat dissipation...112
heterogeneous and unstructured data.................................145
hexadecimal notation..88
hidden patterns and relationships..148
hierarchical addressing and routing infrastructure...............154
High Performance Computing..101
high-level languages...88
highest level of data integrity..91
historical growth trend in HPC...109
historical milieu..133
homogeneous networks..78
HPC community..110
HPC conference in Leipzig..109
HPC performance at a global scale.....................................109
HPC superiority status..108
HPDA...148
Huawei and ZTE..181
human cyber expertise..198
human malice and selfishness..82
hyperbolic but empty statements...195
hypothetical abstraction..16
I/O processes..107
IAB...176
ICA status...58
ICA triad...59

ICANN	176
Icarus complex	34
ICS	54
ICS infrastructure	161
ICT	19
ICT surveillance technologies	181
identity of an ISP	204
idiosyncratic predispositions	135
idle curiosity of our population	65
IETF	176
IGF	177
immature global perception of CYRE	223
impact of Big Data on CYRE	141
impersonal record	201
impersonating a cyber victim	203
impertinence of the American media	62
impertinent agenda of libertarians	59
in-flight Wi-Fi service	204
incipient cyber intrusion	219
incognizance regarding cyber programming	85
incognizant condition	21
incongruous advice	30
increasing on-chip parallelism	113
index of self-importance	139
indolent behavior	57
inevitably flawed human factor	127
inexhaustible supply of IPv6	167
information aggregation	66
INFOSEC	58
infrastructure availability	122
innate pride	34
inordinate degree of infallibility	124
instantaneity	17
interconnect optimization	112
interdependencies among nations	211
interdependent global economy	211
internal cyber security team	198

international manufacturing sources..........189
internationally coordinated legislation..........81
Internet censorship..........179
Internet infrastructure..........172
Internet isolation..........180
Internet utopia..........49
intervening AS..........79
intervening networks..........77
intrusive browser session..........94
invasive Internet filtering system..........181
invention of the transistor..........100
IoE..........155
IoE World Forum..........156
IP address exhaustion..........154
IPSec..........154
IPv6..........153
irrational expectation of privacy..........194
irreparable harm to the US..........70
irresponsible choice of publications..........74
ISOC..........176
ISP..........36
justified belief..........132
Kahn..........174
Kevin Mitnick..........37
Kleinrock..........172
knowledge of CYRE..........19
lacking compliance..........126
large-scale mathematical calculations..........101
latency greater than zero..........107
layer of false assurance..........39
layered cyber security..........30
legacy authority..........177
legal contract with the USG..........69
less complex routing tables..........155
less-resource intensive CPUs..........164
levels of enlightenment..........16
Licklider..........173

limited bandwidth	107
linear causality	18
Linpack performance	105
Linux-based OS	164
logic and abstraction	138
Logically fallacious propositions	51
machine language	89
magnetosphere of the Earth	210
malicious cyber actor	202
malicious impersonator	37
malign cyber code	57
malware	43
man-made reality	207
massive PII compromise cases	127
measurement of computing performance	102
memory management	112
meta-data	47
miniaturization of electronic circuits	100
minimizing manufacturing costs	189
minimum data required	202
minuscule population	11
misinformation fed to the public	70
misrepresentation of cyber	24
misusing the term ontology	13
mitigate the risk of attribution	193
monitoring operation on solar activity	210
multi-core processors	102
multidimensional reality	201
multiplicity of interconnected devices	156
multistakeholder model of governance	176
mutual authentication	154
nascent QC technology	88
National Intelligence Council	162
national security issues	43
national strategies and operations	192
NATO	23
negligent handling of classified information	70

NETMundial	178
NETSEC	58
network data stream	76
network topology	198
networked and networking protocols	29
networked infrastructure	122
new architectural models	102
new mindset	32
no historical antecedents	221
NOAA	210
nominally protected cyber storage	194
non-disclosure agreement	69
non-entity	123
non-intelligent entity	86
nonlinear reality	191
obfuscate the origin	76
obfuscating our lack of prevention	194
obfuscating techniques	218
obfuscating the meaning	205
object code	90
object-oriented high-level computer programming language	86
observable solar event	210
observers	27
obsessive compulsion	145
obsessive use of SNS	48
OCONUS	204
octal notation	88
ODNI	69
omniscience	51
ONCIX	192
ontologia	12
ontological and biological reality	123
ontological conundrum	9
ontological dimension	134
ontological level	10
open and uncensored Internet	184
open-architecture networking model	174

operators...27
OPSEC..58
optimization factor..107
outsourcing model...190
over exposure in the CYRE spectrum...139
P2P..155
packet-switched network concept..173
packets..78
parallel computing..101
parallel cyber universes..11
passive and reactive cyber defense...30
pathetic cyber security record..164
pathological side...35
pattern visualization...148
patterns and relations...148
penetration specialist...206
perception of the cyber reality...130
perennial adversarial global relationships...............................211
performance degradation..192
performance plateau..111
peripheral networking devices...219
permanent remote access...192
persistent access..214
persistent site..93
persistent threat..191
personal concept of CYRE..13
personal cyber exposure...139
personal identifiers..37
personal identity impersonation..37
personal privacy..35
personal safety..35
personal taxonomy..20
personalized encryption tool...205
PETAFLOPS performance level...105
PFLOP level...103
physics and reality..130
picojoules..112

PII	123
pipes and switches	95
plethora of metadata	141
port-scanning	206
position of trust	70
power consumption	112
premeditation of the S-criminal	68
pristine concepts of Internet governance	172
privacy and security requirements	126
private sector espionage	193
proactive cyber defense program	219
proactive monitoring	54
proactive protection	170
probable certainty	133
process of replication of data	143
programming libraries	88
Project Azorian	61
protected cyber storage	194
protection of classified information	70
protection of PII	127
pseudo anonymity	56
pseudo cyber detectives	76
pseudo reality	18
pseudo-Cartesian proclamation	141
psychological aspect	35
psychological compulsion	35
public opinions about CYRE	137
QC reality	26
QKD	26
qualified and experienced cyber SMEs	219
quantum computing	25
quantum cryptography	26
quantum physics	9
quantum teleportation	26
qubits	26
R&D	194
RDBMS	145

readily available Java code	205
recovery plan	83
reducing your cyber footprint	203
relaxing of the rules	44
reliable encryption tools	204
remote cyber entity	42
repressive censorship	179
repressive national governments	179
responsible use of cyber technologies	140
restricted information	69
RFC 1752	153
RFC 2460	153
RFID	162
right to self-defense	195
Robin Sage Experiment	45
rogue access point	206
routing tables	78
scalability limit	106
scalability model	112
SCT	191
security indoctrination	69
security issues plaguing IPv4	153
security patch	132
seemingly chaotic array of information	148
self-interested complacency of journalism	64
self-serving goals	72
Semiotic analysis	84
semiotic indicators	84
sensationalism over the truth	184
sensorial capabilities	136
sensorial manner	138
sequence of connected ASNs	81
serial performance	107
series of very concise instructions	90
set of orderly and coordinated instructions	87
significant challenges	108
simultaneous performance plateau	116

simultaneousness	12
singular machine language instruction	90
small neighborhood of interconnected entities	223
SME	29
SMTP headers	36
SNS	42
SNS as hunting ground	203
SNS playground	44
SNS predators	43
SNS-entity impersonation	45
social dynamics	134
social engineering	37
social media community	148
social media phenomenon	145
social network	75
SORM	181
source code	90
source cyber device	80
source IP address	218
speed of light	17
spiraling conceptual progression	15
spoofed source	82
SQL	145
strong cyber security posture	220
structured and relational databases	145
structured homogeneous data	145
Suanpan	99
sub-atomic particles	9
subjective consciousness	131
subjective reality	131
successful deterrence	30
superiority of the human brain	108
superposition property in a quantum system	87
surveillance and censorship	180
SWPC	212
tailored cyber defenses	30
target for fraud and identity theft	202

target of opportunity	56
targets of opportunity	200
TCO	158
TCP/IP	174
TCP/IP suite of protocols	217
technical aspect of the cyber dimension	216
techno-blindness	49
telecommunications are not cyber	217
teleportation	17
terrorists and cyber criminals	192
the cyber-know	94
The American Black Chamber	60
the BRIC-ITU collusion	179
the cacophony of blogs	140
the CLI sphere of power	93
the cloud model	158
the cyber know-not	94
the cyber know-not	140
the cyber-know	140
the depth and uniqueness of the cyber dimension	222
the end of the cyber age	223
the exascale goal	109
The Freedom House 2013 report	180
the Linux OS	94
the S-criminal	67
three computational systems	104
top-level domains	183
TOP500 List	109
Top5000 HPC systems	104
topologies	27
track the origin of network datagrams	217
tracking methodology	79
trade secrets	195
traditional binary system	87
transcending reality	10
trivial attitude	19
trivial behavior nurtured by SNS	48

true cyber warriors	197
true essence of CYRE	56
trust factor	43
truthful proposition	132
ubiquity	51
UD	60
UDE	58
ultimate danger	35
unauthorized agent	58
unauthorized data extraction	57
unauthorized recipient	71
unauthorized retention	70
unavoidable characteristic	84
unconditional trust	200
underworld of crime or espionage	125
unfounded enthusiasm	108
unique identifier	124
unique identifiers	76
uniqueness of cyber	25
universal entities	14
unjustified personal opinions about CYRE	221
unlawful extraction of PII	126
unprecedented essence of cyber	21
unprecedented reality in history	99
unprotected user login credentials	167
unrealistic and obsolete models	214
unrealistic expectation	108
unrealistic proclivity	124
unscrupulous journalists	63
unveiled reality	16
US-CERT	127
utopian concept of IoE	155
utopians	107
viable proposition	132
Vint Cerf	174
violation of OPSEC protocols	45
violation of the US criminal laws	70

vulnerable to intrusion	125
W3C	176
WAN	173
watch dog of democracy	66
weaponized binaries	53
weaponized cyber techniques	50
web-based email	36
wireless router	206
wonderful and ominous reality	224
world of make-believe	38
world's first computer programmer	86
WSIS	177
WWW	175
"need to know" principle	66
"argumentum ad ignorantiam"	51
"argumentum ex silentio"	51

www.ingramcontent.com/pod-product-compliance
Lightning Source LLC
Chambersburg PA
CBHW020745180526
45163CB00001B/349